WI

PC

OF
GREAT BRITAIN
AND
IRELAND

B.S.B.I. Handbook No. 4

R. D. MEIKLE

ILLUSTRATED BY
VICTORIA GORDON

BOTANICAL SOCIETY OF THE BRITISH ISLES
London
1984

© 1984 Botanical Society of the British Isles

ISBN 0 901158 07 0

Reprinted 1987
Reprinted 1993
Reprinted 1996

Published by the Botanical Society of the British Isles
c/o British Museum (Natural History)
Cromwell Road, London, SW7 5BD
Printed by the Devonshire Press, Torquay, Devon

CONTENTS

Introduction	3
Synopsis of classification and arrangement of species and hybrids	10
Key to *Salix*	13
Key to *Populus*	21
Descriptions and figures	23
References and selected bibliography	186
Glossary	188
Index	193

CORRIGENDA

p. 118	1.6	for *grey-sericeous* read *grey-pubescent*.
p. 122	1.3	for *2–6 cm wide* read *2–6 cm long, 1–3 cm wide*.
p. 129	in heading	for **S. repens L. Salix** read **S. repens L. = Salix.**
p. 132	in heading	for **S. viminalis L. Salix** read **S. viminalis L. = Salix.**
p. 144	1.25	for *many* read *may*.
p. 182	1.8 up	for *Female* read *Male*.

INTRODUCTION

The Family

Salicaceae consists of dioecious, catkin-bearing trees, shrubs and subshrubs, with simple, alternate (rarely opposite or sub-opposite), stipulate leaves, and small flowers in spikes or racemes, each flower subtended by a membranous bract or catkin-scale. The perianth is very reduced, cup-shaped or obliquely cyathiform in *Populus* L.; or of one or more, free or united nectary-scales, or occasionally wanting, in *Salix* L.

The family has an extensive distribution, being represented by indigenous species in almost every part of the world except Australasia, though very much more in evidence in temperate and sub-arctic regions of the northern hemisphere than elsewhere.

The Genera

Most authors have regarded the family as comprising only *Populus* and *Salix*. *Chosenia* Nakai, anemophilous, with pendulous catkins and without apparent nectary-scales, and *Toisusu* Kimura, entomophilous, with lateral nectary-scales in the female flowers, are both so closely similar to *Salix* that subgeneric status would probably be more appropriate for them. *Turanga* (Bunge) Kimura, without a terminal bud, and with unusual polymorphic leaves, is too closely allied to *Populus* to warrant independent generic status.

Allowing for these few aberrations, *Populus* and *Salix* are easily distinguished: *Salix* has a sympodial mode of growth and lacks a terminal bud. *Populus* is almost always monopodial, with a terminal bud. In *Salix* the bud is protected by a single, calyptrate bud-scale; in *Populus* there are several, imbricate scales. The leaves of *Salix* are generally much longer than wide, with short petioles which are not laterally compressed; the leaves of *Populus* are commonly broad-ovate, deltoid or rhomboid, with long, often laterally compressed petioles. The stipules are frequently conspicuous and persistent in *Salix,* but inconspicuous and deciduous in *Populus.* *Salix* catkins are mostly erect or spreading, rarely pendulous, and are usually subtended by several well-developed, often foliaceous, bracts; *Populus* catkins are pendulous, without well-developed bracts. In *Salix* the catkin-scale is generally entire; in *Populus* it is commonly lobed or lacerate. The perianth of *Salix* is reduced to 1 or 2 (rarely more) free or united nectary-scales, that of *Populus* is relatively well-developed, cup-shaped or cyathiform. It is also true to say that *Salix* is primarily entomophilous (insect-pollinated), though it has been shown (Argus in *Canadian Journ. Bot.*, 52:1613–1619(1974)) that a certain amount of *Salix* pollen may be wind-borne; *Populus* is anemophilous (wind-pollinated).

If the two genera are distinguished by many vegetative and floral characters, they are brought together by their capsular, unilocular fruits and small, wholly or almost wholly exalbuminous seeds enveloped in a coma of soft, silky hairs.

Such are the uncertainties in *Salix* taxonomy that estimates of the number of species within the genus vary widely from author to author, from 300 to 500 or more, nor is it possible at present to offer an accurate figure, so many of the names still cited being based upon inadequate specimens and imperfect evidence. *Populus* is a very much smaller genus, most authors agreeing that it includes 30–40 species, chiefly in temperate regions of the northern hemisphere, with concentrations of species in eastern Asia and eastern North America.

Generic Subdivisions

The British and Irish representatives of *Populus* are divisible into 3 relatively well-marked sections: 1. sect. *Populus,* which includes the White Poplar (*P. alba*), the Aspen (*P. tremula*) and their hybrid, the Grey Poplar (*P.* × *canescens*). Members of the section are characterized by their smooth, often lenticellate, bark, lobed or sinuate, often tomentose, leaves, and long-ciliate catkin-scales. 2. sect. *Aegiros* Duby, which includes the Black Poplar (*P. nigra*), a group characterized by fissured bark, glabrous or glabrescent leaves with translucent, cartilaginous, serrate margins and laterally compressed petioles; the catkin-scales are laciniate or lacerate, but not long-ciliate. 3. sect. *Tacamahaca* Spach, the Balsam Poplars, including *P. candicans* and *P. trichocarpa,* with fissured bark, aromatic-resinous buds, and glabrous or glabrescent leaves, often whitish below, but without translucent, cartilaginous margins; the petioles are terete or channelled above, but not laterally compressed, and the catkin-scales are lacerate or laciniate, sometimes shortly pilose, but not long-ciliate. The distinctions between sections *Aegiros* and *Tacamahaca* are somewhat obscured by inter-sectional hybridization.

Salix is less amenable to subdivision. Most authorities accept three subgenera: 1. subgenus *Salix*, the true Willows, including *S. alba*, *S. fragilis*, *S. pentandra* and *S. triandra*; trees and robust shrubs, generally with narrow, acuminate, closely serrate leaves, catkins with pallid, concolorous scales, and nectary-scales 2 (or more) in male flowers and 1 or 2 in females. 2. subgenus *Vetrix* Dumort. (*Caprisalix* Dumort.), the Sallows and Osiers, including many British and Irish species, such as *S. caprea*, *S. viminalis*, *S. purpurea* and *S. phylicifolia*; mostly shrubs with varied leaves, often oblong or obovate, shortly acute or obtuse, with entire or somewhat irregularly toothed or crenate margins; the catkins commonly appear in advance of the leaves on naked twigs; the catkin-scales are often dark-tipped; male and female flowers have one nectary-scale, and the ovaries are frequently pilose. 3. subgenus *Chamaetia* Dumort. including *S. herbacea*, *S. myrsinites* and *S. reticulata*; dwarf arctic or alpine Willows, with the catkins terminal on leafy shoots, the catkin-scales mostly concolorous, often reddish or purplish, and the nectary-scales variously lobed or fused.

These subgenera are useful groupings, but (especially subgenus *Vetrix*) neither homogeneous nor well defined. Attempts to furnish more precise, natural groupings at sectional level have been numerous, each author modifying the concepts of his predecessors, so that sectional analysis appears to be in a permanent state of flux, with almost as many names as there are species. Such are the problems of infrageneric subdivision that one feels the looser and more flexible rank of *series* is better suited to *Salix* than the relatively rigid and exclusive *section*. Furthermore, a satisfactory solution to the problem must involve monographic treatment of the genus as a whole. The last attempt at such an all-embracing survey (2) is now well over 100 years old.

Species

Although, at generic level, there is seldom, if ever, the least difficulty in distinguishing a Poplar from a Willow, the situation with the species is very different, particularly as regards *Salix*. Almost every author dealing with the Willows has commented upon problems of identification, nor can it be said, even when the scope of taxonomic investigation is restricted to the British flora, that all the problems have been satisfactorily solved. As regards *Salix* (and to a lesser degree *Populus*) there are a number of reasons for these taxonomic uncertainties: 1. All Willows are dioecious, with male and female inflorescences on separate individuals. To obtain complete material of any one species it is consequently necessary to collect from at least two separate sources, thereby introducing possibilities of error always present with such composite gatherings. 2. Most Willows and Poplars are proteranthous, producing their catkins at one period of the year, and their foliage at another; to obtain complete or satisfactory material it is usually necessary to revisit collecting sites, and to ensure that catkins and foliage are taken from one and the same individual. Again the possibilities of error are always present, especially where material is selected from extensive, mixed populations, and not from isolated, readily identifiable, trees or bushes. 3. Although Willows are probably no more variable than many other plants not reckoned taxonomically difficult, they can produce very deceptive variants if material for study is collected at the wrong season or from the wrong site. Collectors will often try to accomplish two objectives at one time, by making specimens from material which combines overripe (mostly female) inflorescences with immature foliage, not realizing that, in so doing, they are getting the worst of both worlds, and substantially adding to the possibility of error. Immature foliage of almost any Willow differs from its adult counterpart in size, shape and indumentum, and is often very difficult to identify with certainty. Likewise, collections from very shaded, or very exposed situations are to be avoided; shade-grown Willows often have abnormally large, thin, glabrescent leaves, sometimes with marginal toothing totally unlike that of specimens collected under conditions of normal exposure. Specimens from very exposed sites, or from situations where the plants are affected by drought or starvation, are no better, with growth-habit untypical, leaves small and often abnormally hairy. 4. If asked to explain the problems of *Salix* taxonomy, many taxonomists would point to hybridization as the single most important contributory factor. Willow hybrids certainly outnumber those of any other genus in the British flora, and, in theory at least, it might be possible to contrive a hybrid made up of all the species occurring in these islands, though what is possible by artifice in no way reflects the situation in nature. Indeed our realization that hybrids do occur between many *Salix* species, and are relatively frequent amongst a limited number of species, tends to resolve some of the problems which perplexed earlier writers, who were either unaware of, or blind to, the occurrence of hybrids within the genus. Indeed, knowing that hybridization does occur, the tendency now is to over-emphasize its effect or to put it forward as an easy explanation for difficulties or aberrations which may be attributable to some other cause. A very few hybrids are so prevalent in this country, for example, *Salix caprea* × *S. cinerea*, *S.*

aurita × *S. cinerea* or *S. myrsinifolia* × *S. phylicifolia,* as to obliterate locally distinctions between species which are elsewhere easy to identify. But the great majority of hybrids are either rare, or else represented by uniform, often one-sex, clones, deliberately introduced, or descended from planted specimens, and, with a little experience, almost as easily identifiable as the parents. Such hybrids include *S.* × *sericans, S.* × *calodendron, S.* × *laurina, S.* × *forbyana.* Hybrids known only from one or two records, or of questionable authenticity, are not included in this handbook. Their identification is sometimes less difficult than might be supposed, especially if the collector or recorder looks around and takes stock of possible parentage for the problem plant. Without observations on the composition of Willow populations, the determination of rare and unfamiliar hybrids is often little more than guesswork, as is also the determination of multiple hybrids, involving three or more possible parents.

The taxonomy of *Populus* is scarcely less problematic than that of *Salix,* though for different reasons. The number of *Populus* species native in Britain and Ireland, or naturalized in our area, is so small that identification is relatively easy; but hybrids, of inscrutable complexity, between species and hybrids in sections *Aegiros* and *Tacamahaca,* or between members of these sections, are widely cultivated in parks and gardens, or in commercial plantations, and are regularly augmented by additional, man-made segregates and cultivars, often differing from existing strains only in improved rates of growth, resistance to disease, or difference of habit. A few of the well-established, cultivated *Populus* hybrids are mentioned in this book, but most are excluded.

Terminology

The morphology of Salicaceae, and the structure of Salicaceae flowers is so simple that it calls for little explanation. The terms involved (unlike those used by writers on Orchidaceae) are either widely known, or self-explanatory. The term *catkin-scale* is here used for what some authors call "bracts", and the term *bract* is used for the modified leaves found at the base of a catkin, or along the peduncle bearing the catkin. The distinction between *trunk* and *branches* calls for no explanation, but that between *twigs* and *shoots* is less obvious, and, since it may be important in identification, calls for comment: here the term *twig* is used for the ripened, leafless growths of the preceding season; *shoots* are the leafy, unripened growths of the current season.

Collecting Willows

Since precise identification calls for good material, a few notes on Willow-collecting may be useful. As has been already mentioned, complete specimens, with a very few exceptions, must be collected at two periods of the year, in the spring for catkins, and in the summer or early autumn (July to late September, preferably not earlier or later) for mature foliage. Catkins should be fully developed, but not overripe, and male catkins should be kept apart from females, and separately numbered, not for moral reasons, but because they must normally have come from separate individuals, and may not "belong" together. Since it is essential that catkins and foliage should be taken from the same individual, the source must always be marked, so as to be identifiable on a second visit. Labelling, or slashing the bark with a knife, may be used to identify individual trees or bushes, but, from my own experience, such markings are not wholly satisfactory: labels are easily lost, obliterated or vandalized, and slashes, evident enough when fresh, soon darken and may become almost invisible after a month or two. A neat, simple plan or sketch-map, indicating a particular specimen by reference to some obvious marker—a gate-post, building or large tree—is the best expedient, but it must be exact, and it should be borne in mind that any site looks very different in March or April from its appearance in July or August, so markers should be chosen with care. If it is quite impossible to visit a site more than once, then the visit should be made in summer or early autumn, since foliage specimens are generally much more informative than catkins. In fact it is good policy to collect foliage first and catkins later, so that if a second visit should prove impossible, or if a particular tree or bush cannot be re-located, at least some possibility remains of identifying the single collection. Foliage should be selected from normal growths, not from abnormally vigorous sucker or coppice shoots.

Where it is thought a hybrid is involved, collectors should make sure to collect specimens of likely parents growing in the neighbourhood, though not necessarily in the immediate vicinity, of the suspect. Do not collect sickly specimens from over-shadowed or over-exposed sites, and avoid material infested by fungi or aphids, or heavily gnawed by Sawfly larvae, or any other of the Willow's many enemies. Notes on twig and leaf colour, or on the colour of anthers and catkin-scales are always useful, but notes on size and habit, features rarely commented upon by collectors, are probably most valuable of all.

Keys

An attempt has been made to key out all included species and hybrids, wherever possible on the basis of leaf-characters, since, from experience, the author has found that keys based on catkin-characters are seldom useful. No Willow key (at least none yet devised) will prove infallible, whatever its basis, largely because there are few characters in the genus which are not open to exception, and many which, though obvious enough to the *cognoscenti*, are impossible to describe adequately in words. The illustrations will, it is hoped, provide illumination where the text fails. Willows cannot be completely mastered in the study, though they will, I think, prove amenable in the field to any observer with normal perspicacity and patience.

Acknowledgments

I am grateful to Mr A. O. Chater and Mr A. Newton for invaluable help in editing and seeing the book through the press, to Dr N. K. B. Robson for assistance with proof reading, and to Mr A. C. Jermy for the cover design.

SYNOPSIS OF CLASSIFICATION AND ARRANGEMENT OF SPECIES AND HYBRIDS

SALIX L.
1	S. pentandra L.
2	S. fragilis L. aggr.
2	S. fragilis L. var. fragilis
2a	S. fragilis L. var. furcata Seringe ex Gaudin
2b	S. fragilis L. var. russelliana (Sm.) Koch
2c	S. fragilis L. var. decipiens (Hoffm.) Koch
2×1	S. fragilis L. × S. pentandra L. = S. × meyeriana Rostk. ex Willd.
3	S. alba L. var. alba
3a	S. alba L. var. vitellina (L.) Stokes
3b	S. alba L. var. caerulea (Sm.) Sm.
3×1	S. alba L. × S. pentandra L. = S. × ehrhartiana Sm.
3×2	S. alba L. × S. fragilis L. = S. × rubens Schrank
3a×2	S. alba L. var. vitellina (L.) Stokes × S. fragilis L. = S. × rubens Schrank nothovar. basfordiana (Scaling ex Salter) Meikle forma basfordiana Meikle
3a×2	S. alba L. var. vitellina (L.) Stokes × S. fragilis L. = S. × rubens Schrank nothovar. basfordiana (Scaling ex Salter) Meikle forma sanguinea Meikle
3a×4	S. alba L. var. vitellina (L.) Stokes × S. babylonica L. = S. × sepulcralis Simonk. nothovar. chrysocoma (Dode) Meikle
4×2	S. babylonica L. × S. fragilis L. = S. × pendulina Wenderoth
5	S. triandra L.
5a	S. triandra L. var. hoffmanniana Bab.
5×9	S. triandra L. × S. viminalis L. = S. × mollissima Hoffm. ex Elwert
5×9	S. triandra L. × S. viminalis L. = S. × mollissima Hoffm. ex Elwert var. hippophaifolia (Thuill.) Wimm.
5×9	S. triandra L. × S. viminalis L. = S. × mollissima Hoffm. ex Elwert var. undulata (Ehrh.) Wimm.
6	S. purpurea L.
6×9	S. purpurea L. × S. viminalis L. = S. × rubra Huds.
6×17	S. purpurea L. × S. repens L. = S. × doniana Sm.
13? × 6 × 9	S. × forbyana Sm. = S. ?cinerea L. × S. purpurea L. × S. viminalis L.

7	S. daphnoides Vill.
8	S. acutifolia Willd.
9	S. viminalis L.
10	S. × calodendron Wimm.
11	S. elaeagnos Scop.
12	S. caprea L.
12a	S. caprea L. var. sphacelata (Sm.) Wahlenb.
12×9	S. caprea L. × S. viminalis L. = S. × sericans Tausch ex A. Kerner
12×13	S. caprea L. × S. cinerea L. = S. × reichardtii A. Kerner
13	S. cinerea L.
13a	S. cinerea L. ssp. cinerea
13b	S. cinerea L. ssp. oleifolia Macreight
13×6	S. cinerea L. × S. purpurea L. = S. × sordida A. Kerner
13×9	S. cinerea L. × S. viminalis L. = S. × smithiana Willd.
13×16	S. cinerea L. × S. phylicifolia L. = S. × laurina Sm.
13×17	S. cinerea L. × S. repens L. = S. × subsericea Doell
14	S. aurita L.
14×9	S. aurita L. × S. viminalis L. = S. × fruticosa Doell
14×13	S. aurita L. × S. cinerea L. = S. × multinervis Doell
14×17	S. aurita L. × S. repens L. = S. × ambigua Ehrh.
14×22	S. aurita L. × S. herbacea L. = S. × margarita F. B. White
14×22×17	S. aurita L. × S. herbacea L. × S. repens L. = S. × grahamii Borrer ex Baker
15	S. myrsinifolia Salisb.
16	S. phylicifolia L.
17	S. repens L.
17×9	S. repens L. × S. viminalis L. = S. × friesiana Anderss.
18	S. lapponum L.
19	S. lanata L.
20	S. arbuscula L.
21	S. myrsinites L.
22	S. herbacea L.
22×17	S. herbacea L. × S. repens L. = S. × cernua E. F. Linton
23	S. reticulata L.

POPULUS L.
24	P. alba L.
24×25	P. × canescens (Ait.) Sm. = P. alba L. × P. tremula L.
25	P. tremula L.
26	P. nigra L.
27a	P. deltoides Marsh. × P. nigra L. = P. × canadensis Moench var. serotina (Hartig) Rehder

27b P. nigra L. × P. × canadensis Moench var. serotina (Hartig) Rehder = P. × canadensis Moench var. marilandica (Bosc ex Poir.) Rehder
28 P. candicans Ait.
29 P. trichocarpa Torrey et Gray ex Hooker

KEY TO SALIX

1 Trees and shrubs more than 1 m high (to p. 19)
  2 Leaves linear, lanceolate, oblanceolate or narrowly oblong-elliptic, more than 3 times as long as broad; apex generally acute or acuminate (to p. 16)
    3 Leaves distinctly and closely serrate or serrulate (to p. 15)
      4 Twigs pruinose, often pendulous; catkins conspicuously silky-pilose   **8 S. acutifolia**
      4 Twigs not pruinose; catkins not conspicuously silky-pilose
        5 Stipules persistent, conspicuous; bark smooth, shed in large flakes
          6 Shoots longitudinally ridged or angled, glabrous; leaves glabrous; stipules acute or shortly acuminate, sometimes blunt or rounded
            7 Leaves 4–12(–15) cm long, 1–3(–4) cm wide, often glaucous or glaucescent below; stipules acute or shortly acuminate   **5 S. triandra**
            7 Leaves 2–6(–7) cm long, 1–1.5(–2.5) cm wide, green on both surfaces; stipules generally very blunt or rounded
              5a S. triandra var. **hoffmanniana**
          6 Shoots terete or very obscurely ridged or angled; leaves and shoots at first pubescent or puberulous; stipules with a long, tapering acumen
              5 × 9 S. × mollissima var. **undulata**
        5 Stipules caducous, inconspicuous, often absent; bark rough, fissured, not shed in flakes
          8 Branches pendulous
            4 S. babylonica and hybrids (see p. 59)
          8 Branches erect or spreading
            9 Mature leaves glossy green above (to p. 15)
              10 Twigs clay-coloured (ochraceous); shoots and leaves always glabrous
                2c. S. fragilis var. **decipiens**
              10 Twigs brown or olive-brown, orange-yellow or reddish
                11 Shoots and leaves glabrous; leaf-margins minutely and closely serrulate; male

flowers with (2–)3–4(–5) stamens
            2 × 1 S. × meyeriana
11 Shoots and leaves at first thinly pubescent or ciliate, generally becoming glabrous with age
  12 Twigs orange-yellow or reddish
    13 Leaves 9–15 cm long, with a slender, tapering acumen; catkins more than 6 cm long, often pendulous
        3a × 2 S. × rubens f. basfordiana
    13 Leaves less than 9 cm long, shortly acuminate; catkins less than 6 cm long, erect or spreading
        3a × 2 S. × rubens f. sanguinea
  12 Twigs brown or olive-brown
    14 Leaves irregularly serrate or serrulate with markedly uneven teeth
      15 Leaves generally less than 3 cm wide, with an elongate, attenuate acumen; female flowers with long ovaries much exceeding the subtending catkin scales (male flowers unknown)
          2b S. fragilis var. russelliana
      15 Leaves generally more than 3 cm wide, acumen not very elongate or attenuate; male catkins commonly bifurcate (female flowers unknown)
          2a S. fragilis var. furcata
    14 Leaves regularly serrate or serrulate, teeth not markedly uneven
      16 Leaves minutely serrulate with numerous, close teeth; male flowers generally with 3–4 stamens
          3 × 1 S. × ehrhartiana
      16 Leaves conspicuously serrate, with fewer, more remote teeth; male flowers generally with 2 stamens; female

flowers with ovaries scarcely exceeding subtending catkin-scales
2 S. fragilis var. **fragilis**

9 Mature leaves dull green or sericeous-pubescent above (from p. 13)

17 Twigs yellow or orange
3a S. alba var. **vitellina**

17 Twigs brown or olive

18 Leaves persistently sericeous above
3 S. alba var. **alba**

18 Leaves glabrescent above

19 Leaves remaining thinly sericeous below; tree with ascending branches and a narrow, pyramidal crown
3b S. alba var. **caerulea**

19 Leaves becoming subglabrous on both sides; tree with spreading branches and a broad, rounded crown **3 × 2 S. × rubens**

3 Leaves entire or subentire, sometimes remotely and irregularly toothed, or serrulate only towards apex (from p. 13)

20 Shoots and leaves glabrous; male flower with apparently 1 stamen (actually 2 connate stamens); inner surface of bark yellow; leaves commonly opposite or subopposite
6 S. purpurea

20 Shoots and leaves at first pubescent, pilose or tomentose; male flower with 2 free or partly connate stamens; leaves normally alternate

21 Leaves becoming glabrous or thinly puberulous on both surfaces with age

22 Mature leaves generally quite glabrous, dark lustrous green above, blackening when dried; catkins appearing before the leaves   **?13 × 6 × 9 S. × forbyana**

22 Mature leaves remaining puberulous, at least on the undersurface

23 Catkins appearing with the leaves; catkin-scales yellowish; stipules often persistent
5 × 9 S. × mollissima var. **hippophaifolia**

23 Catkins appearing before the leaves; catkin-scales dark-tipped; stipules caducous or wanting
6 × 9 S. × rubra

21 Leaves remaining pubescent, pilose or tomentose, at least on the undersurface, even at maturity

24 Leaves linear or narrowly linear-lanceolate, the margins almost parallel
  25 Leaves silky below with adpressed silvery hairs; ovary hairy **9 S. viminalis**
  25 Leaves white-tomentose below, not silky; ovary glabrous **11 S. elaeagnos**
24 Leaves lanceolate or narrowly oblong, the margins distinctly convex, not parallel or sub-parallel
  26 Leaves softly pubescent below, with prominent nervation
    27 Stipules conspicuous, persistent; leaf-margins often strongly undulate; wood of peeled twigs with longitudinal striae **14 × 9 S. × fruticosa**
    27 Stipules caducous; leaf-margins not very undulate; wood of peeled twigs smooth
      12 × 9 S. × sericans
  26 Leaves not softly pubescent below; nervation not very prominent
    28 Leaves 4–7 cm long, 0.5–1.5 cm wide, silky below with adpressed hairs
      17 × 9 S × friesiana
    28 Leaves 6–11 cm long, 0.8–2.5 cm wide, thinly adpressed-pubescent below, but not sericeous
      13 × 9 S. × smithiana
2 Leaves ovate, obovate, oblong, elliptic or suborbicular, not more than 3 times as long as broad (from p. 13)
  29 Twigs pruinose; catkins conspicuously silky-pilose; leaves lustrous dark green above, rather regularly serrate, glaucous below **7 S. daphnoides**
  29 Twigs not pruinose
    30 Catkins developing before the leaves on bare twigs (to p. 18)
      31 Leaves regularly acute or shortly acuminate, under-surface ashy-grey; twigs densely pubescent; catkins cylindrical, elongate, 3–5(–7) cm long
        10 S. × calodendron
      31 Leaves not regularly acute or shortly acuminate; catkins not elongate-cylindrical
        32 Stipules conspicuous, persistent; wood of peeled twigs striate
          33 Leaves adpressed-sericeous below, or sometimes on both surfaces **14 × 17 S. × ambigua**

33 Leaves sometimes pubescent below but with spreading, not adpressed-sericeous, hairs
  34 Leaves rugose, with undulate margins, thinly but softly pubescent below; twigs slender, divaricating **14 S. aurita**
  34 Leaves smooth or slightly rugose, with flattish margins; twigs robust, not noticeably divaricating
    35 Apex of leaf often obliquely twisted; mature leaves dull, dark green above, often soft to the touch below **14 × 13 S. × multinervis**
    35 Apex of leaf rarely obliquely twisted
      36 Mature leaves dull, greyish-green above, ashy grey below; twigs pubescent
13a S. cinerea ssp. **cinerea**
      36 Mature leaves lustrous dark green above, greyish or ferrugineous below, with at least some rusty-brown hairs; twigs glabrescent
13b S. cinerea ssp. **oleifolia**
32 Stipules caducous or wanting except on very robust growths
  37 Leaves persistently, softly, pubescent, sericeous or tomentose below
    38 Leaf-margins undulate, irregularly serrate; leaves often very broadly ovate-oblong or suborbicular, green above, densely pubescent below with prominent nervation
12 S. caprea
    38 Leaf-margins flat or recurved, entire or subentire; leaves at first silvery, silky-pubescent on both surfaces, remaining sericeous below even at maturity
      39 Leaves large, 3–7 cm long, 1.5–4.5 cm wide, robust shrubs or small trees
12a S. caprea var. **sphacelata**
      39 Leaves small, less than 5 cm long, 2.5 cm wide, small, slender shrubs
        40 Leaves 1–3.5 cm long, 0.4–2.5 cm wide, dull green above, silvery-sericeous below
17 S. repens
        40 Leaves 2.4–5 cm long, 0.8–2.3 cm wide, bright green above, very thinly sericeous below **13 × 17 S. × subsericea**

37 Leaves becoming thinly pubescent or subglabrous (rarely glabrous) below at maturity
  41 Undersurface of mature leaves with at least some reddish-brown hairs, often distinctly rusty
    13b S. cinerea ssp. oleifolia
  41 Undersurface of mature leaves without reddish-brown hairs
    42 Leaves pubescent below, margins undulate; nervation prominent on leaf-undersurface
      12 × 13 S. × reichardtii
    42 Leaves glabrous or subglabrous on both surfaces at maturity; nervation not very prominent on leaf-undersurface
      43 Leaves conspicuously glaucous below with strongly contrasting upper and lower surfaces; female flower with a well-developed style **13 × 16 S. × laurina**
      43 Leaves not conspicuously glaucous below, nor with strongly contrasting upper and lower surfaces; female flowers with a very short style or stigmas subsessile
        44 Mature leaves glabrous above and below
          6 S. purpurea
        44 Mature leaves generally with a few hairs along midrib and nerves below
          45 Leaves oblanceolate, commonly entire; a low bush 1–1.5 m high
            6 × 17 S. × doniana
          45 Leaves oblong-elliptic or obovate, often with a few marginal teeth; slender bush to 5 m high
            13 × 6 S. × sordida

30 Catkins developing with the leaves (from p. 16)
  46 Mature leaves not blackening when pressed or bruised, rather rigid and subcoriaceous, glossy green above, glabrous or subglabrous on both surfaces; twigs lustrous brown
    47 Leaves paler below, but not glaucous; leaf-margins minutely and closely glandular-serrate; petiole-apex glandular; catkin-scales uniformly pale yellow; male flowers with 5–6 stamens; ovaries glabrous
      1 S. pentandra
    47 Leaves glaucous below; margins entire or irregularly

  serrate; petiole-apex not glandular; catkin-scales dark-tipped; male flowers with 2 stamens; ovaries generally hairy **16 S. phylicifolia**
 46 Mature leaves blackening when pressed or bruised, thin and papery, dull green above, pubescent (rarely glabrescent) below; twigs dull brown or fuscous, generally pubescent; ovaries commonly glabrous or thinly hairy
  15 S. myrsinifolia
1 Small spreading shrubs and creeping subshrubs generally less than 1 m high
 48 Leaf-margins entire or subentire
  49 Stipules conspicuous, persistent
   50 Leaves broadly ovate or suborbicular, thinly woolly or cobwebby above, not adpressed-sericeous below, margins generally flat and entire; catkins large, conspicuous, usually golden-hairy **19 S. lanata**
   50 Leaves oblong or obovate, not woolly or cobwebby above, adpressed-sericeous below, margins usually undulate-serrate; catkins small, inconspicuous, silvery-hairy **14 × 17 S. × ambigua**
  49 Stipules small, caducous or wanting
   51 Leaf-nervation conspicuously reticulate, impressed above, prominent below; petioles long; leaf-undersurface ashy-grey; catkins narrowly cylindrical, long-pedunculate **23 S. reticulata**
   51 Leaf-nervation not conspicuously reticulate; petioles short
    52 Leaves pale grey or whitish below, usually softly tomentose or lanuginose; leaves acute or shortly acuminate, generally lanceolate or narrowly oblong
    18 S. lapponum
    52 Leaves glabrous or thinly pubescent or sericeous below, not pale grey or whitish, nor softly tomentose or lanuginose
     53 Catkins lateral, sessile, developing a little in advance of the foliage; leaf-nervation obscure
     17 S. repens
     53 Catkins terminal on short leafy shoots; leaf-nervation rather prominent
     22 × 17 S. × cernua
48 Leaf-margins serrate or crenate
 54 Creeping subshrubs, generally less than 6 cm high, with

underground rooting stems; catkins terminal on short leafy shoots **22 S. herbacea**
54 Trailing or decumbent, woody shrubs, generally more than 6 cm high; catkins lateral on the twigs
  55 Stipules conspicuous, persistent
    56 Leaves glossy green on both surfaces; catkins large and conspicuous **21 S. myrsinites**
    56 Leaves dull below; catkins small, inconspicuous
      14 × 22 S. × margarita
  55 Stipules small, caducous or absent
    57 Leaves acute, glaucous or glaucescent below, margins minutely and regularly serrulate; ovaries pilose
      20 S. arbuscula
    57 Leaves obtuse or very shortly acute, not glaucous below, margins remotely and irregularly crenate-serrulate; ovaries glabrous or subglabrous
      14 × 22 × 17 S. × grahamii

KEY TO POPULUS

1 Young leaves white or greyish-tomentose on one or both surfaces
  2 Mature leaves (of normal growths) persistently tomentose below; lamina distinctly palmatilobed   **24 P. alba**
  2 Mature leaves (of normal growths) glabrous or glabrescent on both surfaces; lamina coarsely and bluntly sinuate-toothed, but seldom palmatilobed   **24 × 25 P. × canescens**
1 Young and mature leaves glabrous or glabrescent, not tomentose on either surface
  3 Leaves (of normal growths) broadly ovate or suborbicular; margins coarsely and bluntly sinuate-dentate   **25 P. tremula**
  3 Leaves (of normal growths) ovate, ovate-deltoid or cordate; margins shortly serrate or serrulate
    4 Undersurface of mature leaves conspicuously pallid; young leaves strongly balsam-scented
      5 Leaves broadly cordate with an abrupt cuspidate apex and emarginate base   **28 P. candicans**
      5 Leaves ovate, tapering to an acute apex, base cuneate or truncate, rarely subcordate   **29 P. trichocarpa**
    4 Undersurface of mature leaves not conspicuously pallid; young leaves not strongly balsam-scented
      6 Branches erect or sharply ascending; tree conspicuously columnar-fastigiate   **26 P. nigra** var. **italica**
      6 Branches spreading or downcurved, tree not fastigiate
        7 Boughs downcurved; trunk usually with conspicuous swellings or burrs; leaves usually with a truncate or cuneate base; margins often indistinctly serrate   **26 P. nigra**
        7 Boughs spreading or ascending; trunk without burrs; leaves usually with a truncate or cordate base; margins usually distinctly serrate
          8 Branches diverging to form a fan-shaped or goblet-shaped crown; trees male   **27a P. × canadensis** var. **serotina**
          8 Branches spreading to form a broad, rounded crown; trees female   **27b P. × canadensis** var. **marilandica**

DESCRIPTIONS AND FIGURES

1. Salix pentandra L.

Bay Willow

A large shrub or small tree, usually 5–7 m high, exceptionally said to attain 17 m; bark fuscous or greyish, lightly fissured; branches spreading in mature specimens, forming a broad, rounded crown; twigs glabrous, shining as if varnished, generally brown or reddish, supple; buds small, ovoid, pointed, dark brown, glossy and rather viscid. Leaves ovate-elliptic or ovate, sometimes broadly lanceolate or obovate, 5–12 cm long, 2–5 cm wide, base rounded or broadly cuneate, apex acute or cuspidate-acuminate, margins minutely and regularly glandular-serrate; lamina rather coriaceous, dark shining green above, distinctly paler below, obscurely veined; petiole stout, generally less than 1 cm long, bearing several small sessile glands at its apex; stipules minute, ovate, caducous. Catkins appearing with the leaves in late May or June, terminal on short, spreading, leafy, lateral shoots; rhachis and upper part of peduncle densely pubescent. Male catkins shortly cylindrical, 2–5 cm long, about 1–1.5 cm wide, dense-flowered, showy; catkin-scales uniformly pale yellow, oblong, obtuse or subacute, about 2 mm long, 0.8 mm wide, pubescent towards base, glabrous above; stamens usually 5–8 (less often 4 or rarely up to 12); filaments free, clothed with long hairs in the lower half; anthers suborbicular, about 0.5 mm diam., golden-yellow; nectaries 2, the abaxial broad and often crenate at the apex, the adaxial smaller, oblong, truncate or emarginate. Female catkins often rather shorter than the male; ovary shortly stipitate, glabrous, narrowly flask-shaped, attenuate above, about 6 mm long, 1–1.5 mm wide; style short, indistinct; stigmas spreading, shortly 2-cleft or subentire; nectaries 2, similar to, but usually rather smaller than those of the male flowers. Capsule up to 8 mm long, ripening in July.

Frequent by streamsides and on wet ground at low altitudes in northern Britain, but probably not indigenous south of a line drawn from Aberystwyth to Yarmouth, also rare or absent from most of northern Scotland, a surprising fact in view of its wide distribution in northern Scandinavia. Locally frequent in Northern Ireland, but

---

A: leaves × 2/3; B: leaf base × 4; C: ♂ catkin × 2/3; D: detail × 8; E: ♀ catkin × 2/3; F: detail × 8.

Salix pentandra 1

rare and mostly planted in the Republic. Because of its attractive foliage and catkins, *Salix pentandra* appeals to gardeners, and has been planted in many areas far outside its natural range. The male (as with most Willows) is decidedly more ornamental than the female, and consequently more popular in cultivation. In southern Britain the absence of female specimens is often a reliable indication of non-native status.

The Bay Willow is widely distributed in northern and central Europe, but is absent from most of southern Europe and the Mediterranean region.

Many authors compare the fragrance of the glandular leaves and catkins to that of the true Bay, *Laurus nobilis*, but it requires some imagination to detect any of the Bay's spicy aroma in the faint balsamic fragrance of *Salix pentandra*. Linnaeus (*Flora Suecica*, ed.1, 290 (1745)) correctly notes that the leaf serratures exude a gum which marks drying paper with an outline of the leaf in yellow dots. The name "Bay Willow" derives from the name *Salix folio Laureo, seu lato glabro odorato* given to the species by John Ray (*Synopsis,* 216 (1690)).

S. pentandra varies little: forms with narrow, lanceolate leaves are occasionally found, and may be confused with *Salix alba* × *pentandra* or *S. fragilis* × *pentandra*.

2. **Salix fragilis** L. aggr.

Crack Willow

The name Crack Willow is popularly attached to a complex group of tree-willows, consisting of several distinct segregates, all with well-developed trunks, wide-spreading branches, twigs brittle at the point of attachment to the branch, and glossy lanceolate leaves, usually with conspicuous marginal teeth. The distinctions between the variants, set out in detail below, though relatively slight, are so consistent that it is impossible to regard the aggregate as a single, variable unit. The consistency of the infraspecific characteristics is no doubt accentuated, at least in our area, by the fact that most of the variants are clonal, represented by one sex only, and evidently planted where they occur, or derived from planted specimens. Indeed the status of *Salix fragilis* as a whole in Britain and Ireland must be regarded as dubious, since nowhere can it be said to form part of a natural, indigenous plant community. Its standing on the Continent is no less obscure, and made more so by taxonomic uncertainties, as yet unresolved.

2. Salix fragilis L. var. fragilis

Crack Willow

A robust tree, usually 10–15 m high, with a short, thick trunk; bark greyish, coarsely and deeply fissured; branches spreading and forming a broad, rounded crown; twigs at first thinly pubescent, soon glabrous and rather lustrous, terete, olive-brown, brittle at the point of attachment to the branch; buds dorsally compressed, subacute, dull brown. Leaves lanceolate, long-acuminate, 9–15 cm long, 1.5–3 cm wide, dark shining green above, glaucous below, at first adpressed-sericeous, soon glabrous or almost so, base broadly cuneate or rounded, margins coarsely and unevenly glandular-serrate, nervation not very conspicuous, of about 20 sharply upcurved, slender lateral nerves. Petiole rather stout, 5–15 mm long, deeply channelled above, glabrous or thinly hairy, usually having several distinct glands at its junction with the lamina. Stipules generally narrow, acuminate, glandular-serrate, caducous, about 3–8 mm long, 2–3 mm wide at base, exceptionally broad and subfoliaceous (up to 18 mm long and 8 mm wide) in sucker or coppice shoots. Catkins appearing with the leaves in April and May, terminal on short, spreading or suberect, leafy, lateral shoots; rhachis and peduncle densely pubescent. Male catkins cylindrical, 4–6 cm long, 1–1.3 cm wide, rather dense-flowered; catkin scales uniformly pale yellow, oblong, obtuse, about 2 mm long, rather less than 1 mm wide, thinly pilose on the margins and adaxial surface; stamens usually 2, sometimes 3; filaments free or rarely united for a short distance, glabrous above, pilose near the base; anthers oblong, about 0.8 mm long, 0.4 mm wide, golden-yellow; nectaries 2, oblong, blunt or subtruncate, the adaxial usually rather longer than the abaxial. Female catkins similar in general size and shape to male; catkin-scales generally longer, about 3 mm long, and 0.8–1 mm wide, equalling or often slightly exceeding the ovary, soon falling; ovary subsessile or with a very short stalk, flask-shaped, 2.5–3 mm long, about 1 mm wide, tapering from a rounded base, pale green, glabrous; style short, stigmas 2-cleft; nectaries usually 2, shortly oblong, truncate or emarginate. Capsule 4–5 mm long, 2.5 mm wide near base, rather broadly ovoid, generally infertile.

Because of repeated confusion with var. *russelliana* (Sm.) Koch, var. *decipiens* (Hoffm.) Koch and other variants and hybrids, it is almost impossible at present to chart, with any exactitude, the distribution of *Salix fragilis* var. *fragilis* within our area. It is certainly frequent by streamsides and on wet ground throughout the south-eastern counties of England and in the Midlands, but becomes uncommon northwards and westwards, though it extends somewhat sparingly into Cornwall, Wales and southern Scotland. In Ireland, and in most parts of Scotland, it is generally replaced by the introduced var. *russelliana*. Abroad, it has been collected throughout western Europe from Norway to Spain and Portugal, and is found as far east as Roumania, though apparently missing from Greece and western Asia. Its status in our area is no less obscure than its distribution: like *Salix alba*, *Populus nigra* and most of the Elms, it is a tree of deep, moist, lowland soils, and is, in consequence, almost invariably to be found in, or on the borders of, agricultural land, where it is unquestionably often planted, or propagated less deliberately by the frequent use of living (and regenerative) stakes for fencing. It may well be native in many parts of southern England; it is certainly not native in Scotland or Ireland.

Many continental authors regard the tree described above as a hybrid between *S. alba* and *S. fragilis*, identifying the latter with the Willow here named *S. fragilis* L. var. *decipiens* (Hoffm.) Seringe. I feel certain, however, that the Willow described above is the *Salix folio longo latoque splendente, fragilis* of Ray (*Synopsis*, ed.3, 448 (1724)), which must be regarded as the basis of the Linnaean name; Linnaeus himself seems to have more than once altered his stance regarding *S. fragilis*, and the *arbor procera* of the *Species Plantarum* (1753) does not agree well with the species described (and figured) in *Flora Lapponica* (1737). My interpretation of *Salix fragilis* is unquestionably that of J. E. Smith, *English Botany*, 26: t.1807 (1807) and the tree figured here agrees with a specimen sent to me in 1950 by Prof. Olov Hedberg, from Uppsala, "on the banks of 'Fyrisån' near the girls' school" and annotated "probably the type-clone of Linnaeus' description. An old tree". In these

---

A: leaves × 2/3; B: leaf base × 3; C: ♂ catkins × 2/3; D: detail × 8; E: ♀ catkins × 2/3; F: detail × 8.

Salix fragilis var. fragilis 2

circumstances, I am inclined to accept Smith's application of the Linnaean name. The genetic history of *Salix fragilis* L. sec. Sm. is another problem. It could conceivably be the offspring of a cross between *S. alba* L. and *S. decipiens* Hoffm., though the relatively large catkins and coarsely serrate leaves are not what one would expect to find in a hybrid of this parentage. The apparent infertility of most populations of *S. fragilis* (as described above) may be an argument in favour of hybridity, but not one which I would care to use without further systematic observation.

Gilbert-Carter (*13* p. 13) seems to have been the first to note that "the coral-like rootlets which this species sends into the water are red". This is also true for *S. fragilis* vars. *russelliana*, *latifolia* and *decipiens*; the rootlets of *S. alba* are white.

2a. Salix fragilis L. var. furcata Seringe ex Gaudin

Indistinguishable from *S. fragilis* var. *fragilis* in size and habit, but with brighter brown, rather lustrous twigs, and conspicuously broad, coarsely serrate leaves, commonly 3 cm wide, exceptionally up to 4–5 cm wide. The lamina is rather thick and coriaceous, rich glossy green above, and strongly glaucous below. The male catkins are frequently 5–6 cm long, usually more than 1 cm wide, and are often bifurcate; they are dense-flowered, with long catkin-scales, and there are commonly 3 stamens.

This distinctive variety has, in the past, been confused with *S. × alopecuroides* Tausch (*S. fragilis* × *S. triandra*) and with *S. × ehrhartiana* Sm. (*S. alba* × *S. pentandra*), probably because the male flowers commonly have three stamens, though this is by no means unusual in *S. fragilis* var. *fragilis*. It is unquestionably the *Salix fragilior* of Host (*Salix*, 6, t.20; 1828) and *S. fragilis*. L. var. *discolor* f. *latifolia* of Andersson (*1* p. 42). Unfortunately Andersson gave the epithet *latifolia* to two distinct forms of *S. fragilis*, and afterwards (*2* p. 209) united them under *S. fragilis* var. *latifolia*.

Though by no means a common tree in our area, *S. fragilis* var. *furcata* has a wide distribution, occurring sporadically throughout the southern counties of England and extending, though sparingly, as far north as Angus and Easter Ross. Only the male tree has been

A: leaves × 2/3; B: leaf base × 4; C: ♂ catkins × 2/3; D: detail × 8.

Salix fragilis var. furcata 2a

found, and all the evidence points to var. *furcata* being a sport of *S. fragilis*, propagated vegetatively, the whole population probably representing a single clone. On the continent, I have seen specimens from Belgium and Switzerland, but suspect it may be commoner than the records would suggest.

The female specimen figured as *S. fragilior* by Host (*Salix*, t.21) looks very like typical *S. fragilis*.

2b. Salix fragilis L. var. russelliana (Sm.) Koch

Bedford Willow

Similar to *S. fragilis* L. var. *fragilis* in general aspect, but generally a taller tree (the specimens of *S. fragilis* "80 to 90 feet high" mentioned by Bean, *Trees & Shrubs* **4**: 272 (1980) must surely belong to var. *russelliana*), with looser, more graceful, branching, and more slender, less brittle, olive-brown twigs. The leaves are proportionately longer and narrower than those of var. *fragilis*, often 13–15 cm long, and generally about 2–2.5(–3) cm wide, at first very thinly sericeous-pubescent, soon becoming glabrous, the upper surface of the lamina dark, shining green, the lower glaucous, but not very conspicuously so. The leaf apex is long-attenuate, the margins coarsely serrate, with markedly unequal teeth. The glands at the apex of the petiole are commonly subfoliaceous in very robust growths. The female catkins are at first about 4 cm long and 0.8 cm wide, but soon lengthen to 6 cm or more, becoming somewhat pendulous. The catkin-scale, though longer, narrower and more pointed than that of var. *fragilis,* is noticeably shorter than the distinctly pedicellate ovary; it is quickly shed with the development of the catkin. The ovaries are narrow and attenuate, 6–7 mm long and less than 1 mm wide, tapering to a short, but distinct, style.

Quite the commonest and most widespread member of the *Salix fragilis* complex, occurring throughout Britain and Ireland, but always more or less obviously planted. Hardly known on the continent, but so confused with other members of the aggregate that its distribution cannot be accurately plotted.

A: leaves × 2/3; B: leaf base × 4; C: stipule × 2; D: catkins × 2/3; E: mature catkin × 2/3; F: detail × 8.

Salix fragilis var. russelliana 2b

S. fragilis var. *russelliana* was first described by J. E. Smith in 1804 (*26* p. 1045) and rather indifferently figured by Sowerby four years later (*English Botany,* 26: t.1808). It was named *S. russelliana* in honour of John, 6th Duke of Bedford, a very active patron of botany, who, as Smith pointed out, was the first to draw attention to the economic importance of the tree. That the history of *S. russelliana* goes back beyond 1804 is clearly evident from comments made by Smith in *English Botany,* and from a letter dated 11th November 1804 from the Duke of Bedford (published in *Memoir and Correspondence of the late Sir James Edward Smith,* 2: 111 (1832)) where it is stated that "my grandfather [4th Duke of Bedford 1710–1771] introduced the same Willow into his county forty or fifty years ago". Twenty five years later, in the introduction (p.v.) to the *Salictum Woburnense* (1829), the Duke gave rather a different account of the origin of *S. russelliana,* saying that it "was originally introduced into Woburn plantations from Leicestershire, sent to the late Duke of Bedford [Francis, 5th Duke of Bedford] by the celebrated Mr Bakewell [Robert Bakewell (1725–1795), a noted agriculturist of Dishley, Leics.]; it then went by the name of the Dishley or Leicestershire Willow". William Withering Jr. (*Systematic Arrangement,* ed.5, 2: 61 (1812)) says it was the Lichfield Willow, so beloved by Dr Samuel Johnson (see *Gentleman's Magazine* June 1785), but if the Willow depicted by Withering is a true representation of Johnson's tree, then it was certainly not var. *russelliana*. This famous specimen, after suffering a number of vicissitudes, was finally blown down, at a reputed age of 130 years, in September 1829, but according to Loudon (*17* p. 1520) "After the tree was blown down, Mr Holmes, a coachmaker residing in Lichfield, and proprietor of the ground on which Johnson's Willow stood, regretting that there was no young tree to plant in its stead, recollected that, the year before, a large branch had been blown down, part of which had been used as peasticks in his garden, and examined these, to see if any of them had taken root. Finding that one had, he had it removed to the site of the old tree, and planted there in fresh soil; a band of music and a number of persons attending its removal, and a dinner being given afterwards by Mr Holmes to his friends and the admirers of Johnson. The young tree is, at present, in a flourishing state, and 20 ft. high". It would be interesting to know if this offspring from Johnson's Willow still survives, and if it is var. *russelliana*. Withering's son says that Johnson's Willow was "situated by the foot-path leading direct from the Minster to Stowe". He also states that *S. fragilis* var. *russelliana* is "known in Ireland by the name of *Gorgomel Sallow,* as the

Bishop of Dromore informs the Rev S. Dickenson. It is not indigenous there, but had been brought from Holland". This would be interesting if it were true, but Wade (*29* p. 359) calls *S. acuminata* Sm. [*S.* × *calodendron* Wimm.] the "Jargomel sallow", and one suspects that this, rather than *S. fragilis* var. *russelliana,* is also the "Gorgomel".

In view of its fame, it is surprising how quickly *S. fragilis* var. *russelliana* came to be confused with other *fragilis* segregates. Andersson (*1* p. 43) considered it synonymous with *S.* × *viridis* Fries, one of the many names given to the hybrid *S. alba* × *fragilis,* and was followed by Boswell-Syme, who, in the third edition of *English Botany* (t.1308) re-issued Sowerby's illustration of *S. russelliana* under the name *S. viridis*. Buchanan White (*31*) further fogged the issue by mixing two distinct segregates in his var. *britannica:* male var. *britannica* appears to be our *S. fragilis* var. *fragilis* or perhaps *S. alba* × *fragilis*; female var. *britannica* is, on Buchanan White's own admission, *S. fragilis* var. *russelliana*. Buchanan White's treatment of the *alba—fragilis* complex is so intricate and obscure that Linton (*16*) not surprisingly acknowledged defeat by uniting both var. *russelliana* and var. *britannica* under aggregate *S. fragilis*. Moss (*18*) regarded *russelliana* as "a particular form or segregate" of *S. alba* × *fragilis,* and concluded that this "particular segregate or mutant × *S. russelliana* has been lost sight of; but its alleged economic importance makes its rediscovery desirable". There can be little doubt that var. *russelliana,* like var. *furcata,* is a straightforward sport of *S. fragilis,* almost certainly a single clone, propagated vegetatively. As to its economic value, modern commerce has little use for the tree even if, as Smith says, its bark contains "more of the tanning principle than any other tree of this country, except the oak".

There have been frequent reports of male var. *russelliana* since Johnston first recorded it from New-water-haugh plantation, in the *Flora of Berwick-upon-Tweed* (1829); most of these records turn out to be var. *fragilis* or *S. alba* × *fragilis*; to date I have not seen male specimens which can confidently be assigned to the Bedford Willow.

2c. **Salix fragilis** L. var. **decipiens** (Hoffm.) Koch

White Welsh Willow

Generally a robust, twiggy bush 5–7 m high, but occasionally developing into a tree 10–15 m high, with coarsely fissured grey-brown bark, like that of *S. fragilis* var. *fragilis,* and with similar brittle twigs. The year-old twigs are pale ochre-coloured and distinctly lustrous, quite unlike those of any other *fragilis* segregate, and the fresh, unripened shoots are often stained crimson. The leaves are wholly glabrous, proportionately shorter and broader than those of var. *fragilis* or var. *russelliana,* rarely exceeding 9 cm in length, but frequently 2–3 cm wide, bright glossy green above, conspicuously glaucous below, with a short acuminate apex and coarse marginal teeth. The male catkins, which are solely found in our area, are generally less than 3 cm long and 0.7 cm wide, and have a puny, under-nourished appearance, standing erect on short, condensed, leafy, lateral shoots. The rhachis is shortly and thinly pubescent. Each flower has a short, blunt, rather densely pilose catkin-scale, usually about 6 mm long and almost as wide, but often not much shorter than the rather thick filaments, which are glabrous except at the base. The anthers are broadly oblong or suborbicular. The flowers are 2-or sometimes 3-stamened, both sorts occurring in each individual catkin. There are 2 oblong nectary-scales, the abaxial usually larger and longer than the adaxial. (The female catkins are often a little larger than the male, with similar catkin-scales, and with a shortly pedicellate, tapering ovary, about 2–2.5 mm long and less than 1 mm wide).

As mentioned under *Salix fragilis* var. *fragilis,* many continental authors regard this as a species, and, with *S. alba,* a parent of the *S. fragilis* of Smith and other British authors. Nomenclaturally, I am satisfied that it is not the *Salix fragilis* of Linnaeus, and, were I to adopt the taxonomy of continental authors, would feel obliged to continue to name it *S. decipiens* Hoffm., and to use the name *S. fragilis* for the hybrid *S. alba* × *S. decipiens.* However, I am by no means certain the *S. fragilis* L. sec. Sm. is a hybrid, nor am I convinced that *S. decipiens* is more than a very distinctive

---

A: leaves × 2/3; B: leaf base × 4; C: stipule × 4; D: catkins × 2/3; E,F: detail × 12.

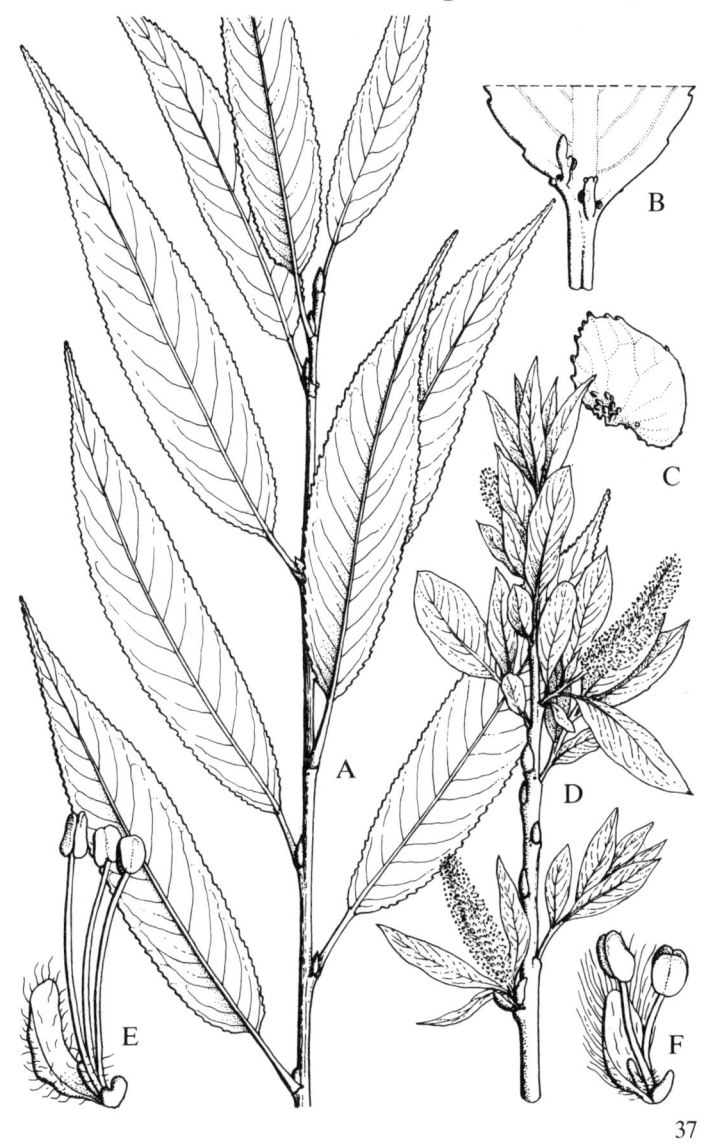

Salix fragilis var. decipiens 2c

variant—perhaps even a subspecies—of the latter. It has a wide distribution in Europe, but, as in our area, the male tree is alone found in many areas, though I have seen male and female specimens from Germany and Austria, but, again, often from gardens or Osier holts and almost certainly subspontaneous.

In our area it was first recorded by Winch (*Botanists' Guide,* 1: 90 (1805)) from "the Banks of Lyne, Team and Derwent", and has since been noted in most counties, though seldom in quantity, and invariably as an introduction.

Buchanan-White's opinion that *S. decipiens* is a hybrid (*S. fragilis* × *S. triandra*) has been largely (but perhaps unwarrantably) rejected; his comment (*31* p. 351) that the female tree occurs in Britain is not borne out by the study of herbarium material, and may be based on a misidentification; he does not cite any precise locality for female specimens.

Smith (*English Botany,* t.1937 (1808)) was the first to call *S. fragilis* var. *decipiens* the White Welsh Willow, remarking that James Crowe had sent him specimens, under this name (or, more exactly, as the "White Welsh Osier") from Osier beds in Norfolk and Cambridgeshire. Elwes and Henry (*8* p. 1756) identified it with "*S. cardinalis*" and the "Belgian Red Willow"; this name is, however, more usually applied to red-barked forms of *S. alba* var. *vitellina,* or to the hybrids between this and *S. fragilis.*

2 × 1.   Salix fragilis L. × S. pentandra L. = Salix × meyeriana Rostk. ex Willd.

A large bush or small tree, generally less than 15 m high, with fissured bark like *S. fragilis,* and glabrous, shining brown twigs, like those of *S. pentandra.* The leaves also resemble those of *S. pentandra,* being coriaceous in texture, dark shining green above, paler below, 5–12 cm long, 1.5–4 cm wide, tapering to a rather long acumen in fully developed specimens, the margins minutely and regularly glandular-serrate. Catkins longer and more narrowly cylindrical than in *S. pentandra,* 4–5 cm long, about 0.8 cm wide,

---

S. × *ehrhartiana* A: leaves × 2/3; B: ♂ catkin × 2/3; C: detail × 8;
S. × *meyeriana* D: leaf × 2/3; E: ♂ catkin × 2/3; F: detail × 8; G: ♀ catkin × 2/3; H: detail × 8.

Salix × ehrhartiana 3 × 1
Salix × meyeriana 2 × 1

appearing with the leaves in May. Male and female trees are both found in our area, the former with (2–)3–4(–5) stamens in each flower, the latter with a narrowly flask-shaped, glabrous ovary about 4 mm long, and considerably exceeding the oblong, obtuse, thinly hairy catkin-scale.

Salix × *meyeriana* occurs sporadically over much of England, and has been recorded from three Irish vice-counties, though one of these records (from Kildare) is erroneous, being a misidentification of a *S. fragilis* variant. The hybrid has not yet been satisfactorily recorded from Scotland. A recent study (Maxted & Trueman in *Watsonia*, 14: 337–346 (1983)) of Shropshire populations of *S.* × *meyeriana*, suggests that here, as elsewhere, the hybrid has been introduced, and is not spontaneous; all the individuals examined were found to be female.

It is usually found as a solitary tree, or in small quantity, in moist thickets and by river margins at low altitudes, and has the appearance of being deliberately introduced in all its stations, presumably as an ornamental, since it is unlikely to have any value as a basket-willow. *S.* × *meyeriana* also occurs widely, but sparingly, in central and northern Europe.

3. Salix alba L. var. alba

White Willow

A tall, graceful tree, generally 10–25 m high, but specimens 30 m high have been recorded; trunk usually well developed; bark greyish-brown, deeply fissured; principal boughs and branches ascending at a sharp (30°–50°) angle and often forming a narrow, pointed or truncate crown, or a number of pointed "turrets" though the ultimate divisions may be spreading or even somewhat pendulous; twigs at first densely pubescent with adpressed, silky hairs, becoming glabrous and rather lustrous brown or olive in their second year, usually supple; buds small, ovoid, pointed, dark brown, generally covered with adpressed pubescence. Leaves

---

A: leaves × 2/3; B: ♂ catkins × 2/3; C: detail × 8; D: ♀ catkins × 2/3; E: detail × 8.

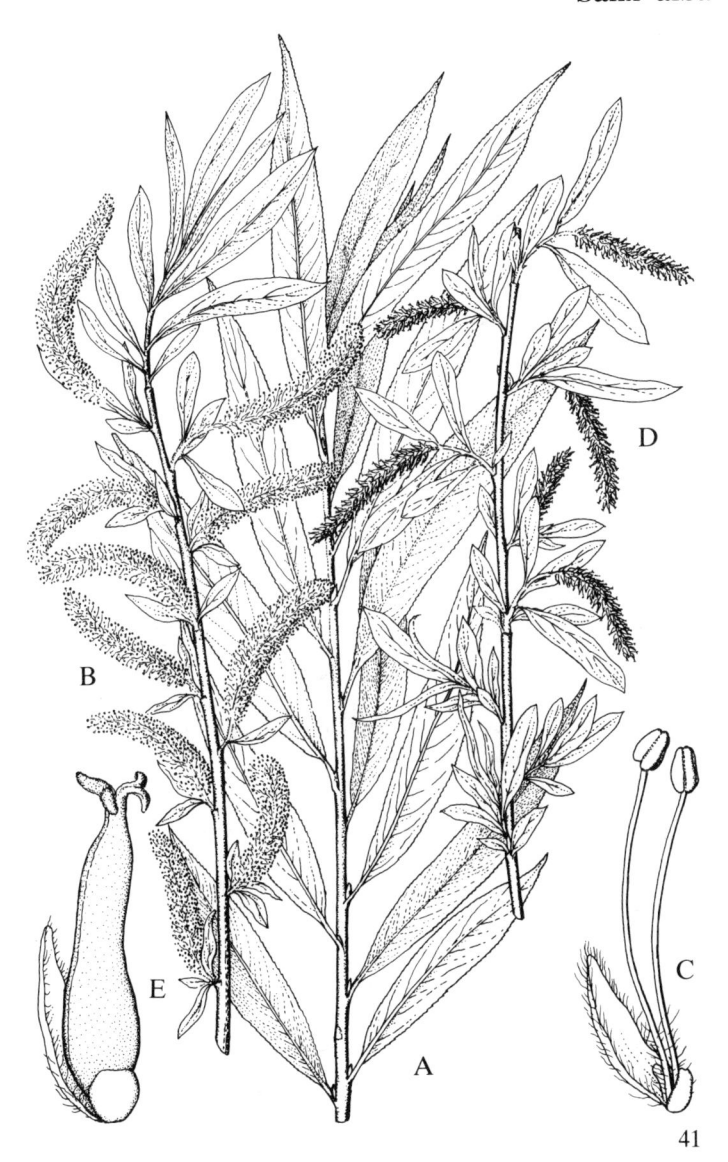

Salix alba

lanceolate-acuminate, 5–10(–12) cm long, 0.5–1.5 cm wide, base narrowly cuneate, apex tapering to a slender acumen, margins minutely and rather regularly serrate; lamina at first silvery-grey with dense adpressed pubescence, the upper surface generally becoming dull green and glabrescent with age, the under surface remaining greyish or glaucescent; venation rather obscure; petiole short, generally less than 1 cm long, pubescent, usually with 4–6 or more small, dark, sessile glands towards the apex; stipules caducous, very narrow, almost subulate, usually less than 5 mm long and 1.5 mm wide at base, margins minutely glandular-serrate. Catkins appearing with the leaves in late April or early May, terminal on very short, leafy, lateral shoots, spreading or suberect; rhachis and peduncle densely pubescent. Male catkins elongate-cylindrical, 4–5 cm long, about 0.8 cm wide, rather dense-flowered; catkin-scales uniformly pale yellow, oblong, tapering to a blunt apex, 2–3 mm long, 1.5 mm wide, pubescent towards base, glabrous above except at margins; stamens 2; filaments free, glabrous or thinly pubescent especially in the lower half; anthers yellow, broadly oblong, about 0.4 mm long, 0.3 mm wide; nectaries 2, oblong, truncate or shortly emarginate at apex, the adaxial usually broader and rather longer than the abaxial. Female catkin rather shorter and narrower than the male, about 3–4 cm long, 0.4 cm diam.; ovary subsessile or very shortly stipitate, glabrous, narrowly flask-shaped, about 3–4 mm long, 1 mm wide; style very short and indistinct, stigmas spreading, 2-cleft, nectary 1, adaxial, broadly oblong with a truncate apex. Capsule broadly and rather shortly flask-shaped, about 4 mm long, ripening in July.

Locally common by rivers and streams in lowland areas of eastern Britain and Ireland, generally becoming scarcer westwards, especially in Wales and Scotland. *Salix alba* is so widely planted that the limits of its natural distribution are probably no longer ascertainable. It is widespread in Europe, extending to the Mediterranean region and eastwards to Central Asia, and has a better claim to being considered indigenous in our area than any of the varieties of *S. fragilis*.

The silveriness of the young foliage makes it an easy tree to recognise even at a distance.

In addition to typical *S. alba* (var. *alba*) described above, two well-marked varieties (var. *vitellina*, var. *caerulea*) are also extensively cultivated in our area, and hybrids also occur between *S. alba* and *S. pentandra*, *S. fragilis*, and *S. babylonica*.

3a. Salix alba L. var. vitellina (L.) Stokes

Golden Willow

Similar to *S. alba* var. *alba* in habit, but usually rather less robust, with the year-old twigs bright yellow or orange, and very conspicuous during the winter months. Leaves as in var. *alba*, but generally more glabrescent, with the upper surface of the lamina soon becoming bright, rather lustrous green. Male catkins resembling those of var. *alba*, but with the catkin-scales usually rather longer, commonly 3–3.5 mm long, and more thinly pubescent; the greater length of the catkin-scale is even more noticeable in the female flower, for the ovary is generally smaller and shorter than in var. *alba*, and is commonly equalled, or sometimes considerably exceeded by the subtending scale, giving the catkin a characteristic lean and rather ragged appearance.

Salix alba var. *vitellina*, though long established in Britain and Ireland, has no claim to be considered a native. It was formerly used in basket-work, and is still commonly seen about Osier grounds, though its chief value nowadays is as an ornamental, and, apart from *S. babylonica* hybrids, it is probably the most popular of Willows in gardens, where it is sometimes heavily pruned each year to make it produce long, bright-barked growths for winter decoration. *S. alba* L. var. *britzensis* Spaeth (*S. alba* L. var. *chermesina* hort.) is a cultivar of var. *vitellina* with exceptionally brilliant, red or orange-red twigs.

The origin of the Golden Willow is obscure. In view of the fact that it is distinguished not only by the colour of the twigs, but by leaf and catkin characters, one suspects it is not just a simple mutant of typical *S. alba*, but more probably one of the many geographical races of this species. At the eastern and northern limits of its range *S. alba* is represented by variants often very far removed in general appearance from the tree regarded as typical in lowland Britain and Ireland.

3b. **Salix alba** L. var **caerulea** (Sm.) Sm.

Cricket-bat Willow

Resembles *S. alba* L. var. *alba* in general appearance, with a pyramidal crown, ascending branches, but with rather larger, broader leaves, often 10–11 cm long and 1.5–2 cm wide. The young leaves are densely adpressed-pubescent with the silvery hairs characteristic of *S. alba* var. *alba*, but this indumentum is soon shed, and the fully-developed foliage is dull bluish-green above, and glaucescent below, subglabrous, or very thinly and sparingly pubescent, often with rather conspicuously toothed margins. In other respects *S. alba* var. *caerulea* is virtually indistinguishable from var. *alba*.

The original Cricket-bat Willow is generally supposed to have been found by James Crowe in the parish of Eriswell, Suffolk in 1803, but Smith (*English Botany*, 34: t.2431 (1812)) mentions a ten-year old tree, planted at Heatherset, Norfolk and blown down in 1800, so it would appear that the variety was already known, at least locally, by 1790. It soon gained a reputation for vigorous growth, Smith (*English Botany, loc. cit.*) reports that the Heatherset tree had become, in ten years, "thirty-five feet high, and five feet two inches in girth". Elwes and Henry (*8* p. 1767) record a 53 year-old tree cut down at Boreham, Essex in 1899 which was 101 feet in height, 5 feet 9 inches in diameter, weighed upwards of 11 tons, and provided wood for 1179 cricket-bats. It is often said that only the female tree is known, but both sexes occur, though as the female is alone considered of first-class quality for bats, the male is seldom planted.

Smith first described the tree as a distinct species (*English Botany*, 34: t.2431 (1812)) but subsequently reduced it to varietal rank, remarking that the only distinctive character he could find was the fact that the "underside of the *leaves* loses, at an early period, most of its silky hairs". The exceptional vigour of the tree, together with the size, serrature and glabrescence of the leaves, suggest that it may be one of a whole series of hybrids which connects *Salix alba* and *S. fragilis*, a suggestion further supported by the considerable

---

A: leaves × 2/3; B: leaf base × 3; C: ♂ catkins × 2/3; D: detail × 12; E: ♀ catkins × 2/3; F: detail × 8.

Salix alba var. **vitellina** 3a

variation to be found in *S. alba* var. *caerulea*, even amongst cultivated specimens. In all essentials, however, the tree is allied to *S. alba* rather than to *S. fragilis*.

The Cricket-bat Willow is common and widespread in southeastern England, from East Anglia to Hampshire; around London it commonly replaces typical *S. alba*. Perhaps misled by the common occurrence of var. *caerulea* in the London area, John Fraser (*11* p. 719) came to regard it as genuine *S. alba*, and distinguished typical var. *alba* as *S. alba* L. var. *stenophylla*.

S. alba var. *caerulea* is seldom mentioned by continental botanists, nor is it possible to glean any precise information about its distribution outside Britain.

3 × 1. Salix alba L. × S. pentandra L. = Salix × ehrhartiana Sm.

Generally a small tree, 10–15 m high, but occasionally up to 25 m, with deeply fissured bark and sublustrous (not highly polished) brown or olive-brown, glabrous twigs. Leaves exactly intermediate between those of the parents, lanceolate or narrowly oblong-elliptic in outline, 6–10 cm long, 1.3–2.5 cm wide, at first thinly adpressed-pubescent, especially on the upper surface, soon becoming glabrous and shining green, slightly paler below, acumen long and tapering, margins minutely and regularly serrate. The male catkins appear with the leaves in late April and early May, and are much longer and more slender than those of *S. pentandra*, usually 3–6 cm long, and less than 1 cm wide, with a relatively short peduncle; the catkin-scales are pale and thinly pubescent as in *S. alba*, with (2–)3–4)–5) stamens in each flower. The female tree does not appear to grow in the British Isles, nor does it seem to be common abroad. To judge from Swedish specimens, the female catkins are long and cylindrical, as in *S. alba*, with similar, shortly flask-shaped ovaries.

An uncommon tree, ornamental in leaf and flower, and probably always planted or occurring as a fairly obvious escape from

---

A: leaves × 2/3; B: ♂ catkins × 2/3; C: detail × 12; D: ♀ catkins × 2/3; E: detail × 12.

Salix alba var. caerulea

cultivation. It is found here and there in Surrey, Essex, Kent, Herts, and Cambridge, and probably elsewhere in S.E. England, also in Cumbria, where it is locally frequent, especially in the Eden valley area. The Scottish records are unconfirmed. *Salix* × *ehrhartiana* has a limited distribution in central and northern Europe.

3 × 2. Salix alba L. × S. fragilis L. = Salix × rubens Schrank

A tall tree, 20–25 m high, with a thick trunk and coarsely fissured bark; branches usually spreading and forming a broad rounded crown; twigs terete, at first thinly pubescent, soon glabrous and dull or somewhat lustrous brown or olive-brown. Leaves lanceolate or linear-lanceolate, generally long-acuminate, at first rather densely appressed-sericeous, soon becoming glabrous, but not shining green above, glaucous below, and often remaining thinly appressed-pilose; margins minutely, but sometimes rather irregularly, serrate. Catkins spreading or suberect, rather narrowly cylindrical, 3.5–6 cm long, usually less than 0.8 cm wide; rhachis and peduncle densely pubescent. Male catkins often rather lax-flowered, with the blunt, pilose catkin-scales generally much shorter than the stamens; ovary glabrous, subsessile or shortly pedicellate, generally flask-shaped.

Under the name *Salix* × *rubens* one finds a wide range of variants, at one extreme merging with *S. alba* L. var. *caerulea* (Sm.) Sm., and at the other difficult to distinguish from *S. fragilis* L. The opinion, widely held on the continent, that *S. fragilis* of British authors is but one of a series of hybrids between *S. alba* and true *S. fragilis* (*S. decipiens* Hoffm.), would appear to be borne out by the observation that the two supposed parents have distinct characteristics seemingly combined in *S. fragilis* of British authors and *S. alba* × *fragilis*. The arguments against the continental hypothesis are set out under *S. fragilis*; it may be added that the difficulties in separating *S. alba* and *S. fragilis* are further magnified when studies of the group are broadened to include western and central Asia.

S. × *rubens* Schrank is widely distributed in Britain and Ireland, but usually as a planted tree or a fairly obvious derivative of cultivation. It was, until recently, often identified with *S. viridis* Fries, but this strange Willow, which, to the best of my knowledge, is not found in our area, has very glabrescent shining leaves, and long, erect, cylindrical catkins, the males dense-flowered, with very pilose scales, the females with short, subsessile ovaries.

a. × 2. **Salix alba** L. var. **vitellina** (L.) Stokes × **S. fragilis** L. = **Salix × rubens** Schrank nothovar. **basfordiana** (Scaling ex Salter) Meikle

This hybrid is represented in Britain by two distinct segregates:

) **Salix × rubens** Schrank nothovar. **basfordiana** (Scaling ex Salter) Meikle forma **basfordiana** Meikle

A vigorous, broad-crowned tree, usually 10–15 m high, with a short trunk and widely spreading branches; twigs rather slender and flexible, glabrous and lustrous orange-yellow, buds usually rather elongate and acute, pale greenish-yellow. Leaves narrowly lanceolate, long-acuminate, 9–15 cm long, generally less than 2 cm wide, bright shining green above, glaucous below, at first sparsely pubescent or ciliate, soon quite glabrous, base rather narrowly cuneate, margin somewhat unevenly, but not coarsely serrate, nervation not very conspicuous, with numerous sharply upcurved, slender lateral nerves. Petiole rather stout, glabrous, deeply canaliculate above, 1–1.8 cm long, with 1–3 inconspicuous glands near its apex. Stipules narrow, caudate-acuminate, glandular-serrate, very caducous, usually about 3–8 mm long, 2–3 mm wide at base. Catkins appearing with the leaves in April and May, terminal on short, spreading, leafy lateral shoots; rhachis and peduncle densely pubescent. Male catkins narrowly cylindrical, 7–8 cm long, 1–1.3 mm wide when fully developed, spreading or pendulous; catkin-scales uniformly pale yellow, oblong, about 2–2.5 mm long, rather less than 1 mm wide, thinly pilose especially along the margins; stamens usually 2(–4); filaments free, glabrous above, shortly pubescent near base, 3–4 mm long, usually much exceeding the subtending scale; anthers broadly oblong, about 1 mm long and almost as wide; nectaries as in *S. fragilis* L. Female catkins very long and pendulous, commonly exceeding 8 cm, exceptionally 11–12 cm long, catkin-scales generally larger, and proportionately narrower than in the males, sometimes as much as 4 mm long, minutely pubescent towards base or all over, caducous; ovary subsessile or shortly stalked, flask-shaped, tapering to a narrow apex, pale green, glabrous; style about 0.5 mm long, stigmas 2, spreading, about 0.5 mm long, often deeply bifid. Developed capsule not seen.

This handsome Willow is said to have arisen sometime before 1870 as a seedling in the nursery of the willow-grower, Mr William Scaling, of Basford, Notts. Flowering material of the male tree had been added to Leefe's herbarium by 1874, but the name *Salix basfordiana,* given to the Willow by Scaling, was not validly published until March 1882 (*Gardeners' Chronicle.* n.s., 17: 298) when James Salter described and figured the male plant, unfortunately at the same time confusing it with a female plant of a very closely allied hybrid (here designated *S.* × *rubens* Schrank nothovar. *basfordiana* (Scaling ex Salter) Meikle f. *sanguinea* Meikle) which Scaling had distributed in 1871 under the invalid name *S. sanguinea.*

S. × *rubens* nothovar. *basfordiana*, though nowhere common, is widely distributed and often subspontaneous in counties adjacent to London. Specimens have also been seen from areas as far apart as the Scillies, Norfolk, Lancashire and Kintyre, and one suspects that it is now to be found in most parts of England, though evidently rare in Scotland, and, as yet, unrecorded from Ireland. It is often confused with *S. alba* L. var. *vitellina* (L.) Stokes, from which it is at once distinguished by its lustrous leaves and long, pendulous catkins. The orange-yellow twigs are very conspicuous during the winter months.

A: leaves × 2/3; B: ♂ catkins × 2/3; C: detail × 8; D: ♀ catkins × 2/3; E: detail × 8.

Salix × rubens forma **basfordiana** 3a × 2 (a)

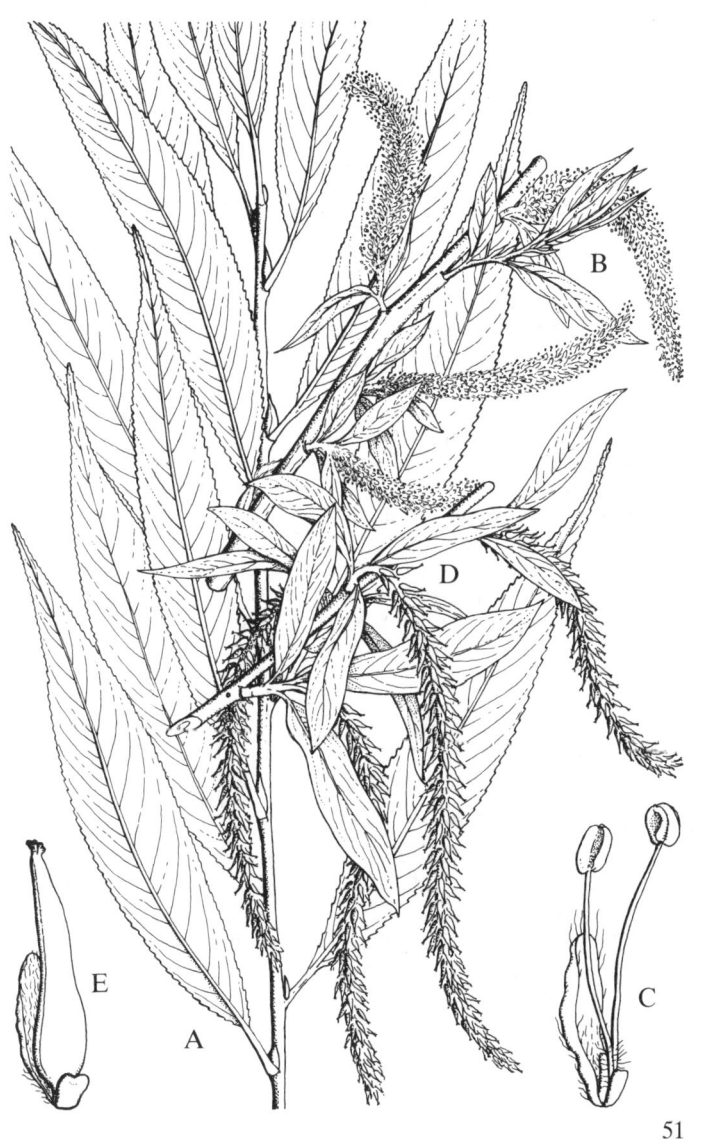

b.) **Salix** × **rubens** Schrank nothovar. **basfordiana** (Scaling ex Salter) Meikle forma **sanguinea** Meikle

---

Superficially similar to *S.* × *rubens* nothovar. *basfordiana* f. *basfordiana,* but a lower growing, less vigorous tree, often scarcely more than a bush, with darker red twigs. The leaves approach those of *S. alba* L. var. *vitellina* (L.) Stokes, seldom exceeding 8 cm in length or 1.5 cm in width, with an acuminate apex, narrowly cuneate base, and very shortly serrate margins. The female catkins are spreading or suberect, 3–4 cm long, rarely more than 0.7 cm wide, with a sparsely pubescent peduncle and rhachis, and narrowly oblong, glabrescent, catkin-scales about 3 mm long and less than 1 mm wide. The narrowly flask-shaped ovaries are only a little longer than the subtending scales, and have a short style with 2 blunt, bifid stigmas. Authentic male specimens have not been seen.

Salix × *rubens* nothovar. *basfordiana* f. *sanguinea* is less often planted than f. *basfordiana,* probably because the whole tree is less graceful and the twigs less colourful. It is occasionally seen, mostly as an obvious introduction, about London and the home counties. Scaling recorded that it was first found in the forests of the Ardennes, but it does not appear to be known to continental salicologists. Like f. *basfordiana,* it can be distinguished from all forms of *S. alba* L. var. *vitellina* (L.) Stokes by its shining green leaves.

3a × 4. **Salix alba** L. var. **vitellina** (L.) Stokes × **S. babylonica** L. = **Salix** × **sepulcralis** Simonk. nothovar. **chrysocoma** (Dode) Meikle

---

A "weeping" tree, usually not much more than 12 m high; bark greyish-brown, deeply and coarsely fissured; twigs very slender, at first thinly subadpressed-pubescent, soon becoming glabrous and golden- or greenish-yellow; buds narrowly ovoid, acuminate, rich brown, glabrous or thinly hairy, especially about the apex. Leaves narrowly lanceolate, acuminate, 7–12 cm long, 0.7–1.8 cm wide,

---

A: leaves × 2/3; B: leaf base × 4; C: catkins × 2/3; D: detail × 8.

Salix × rubens forma sanguinea 3a × 2 (b)

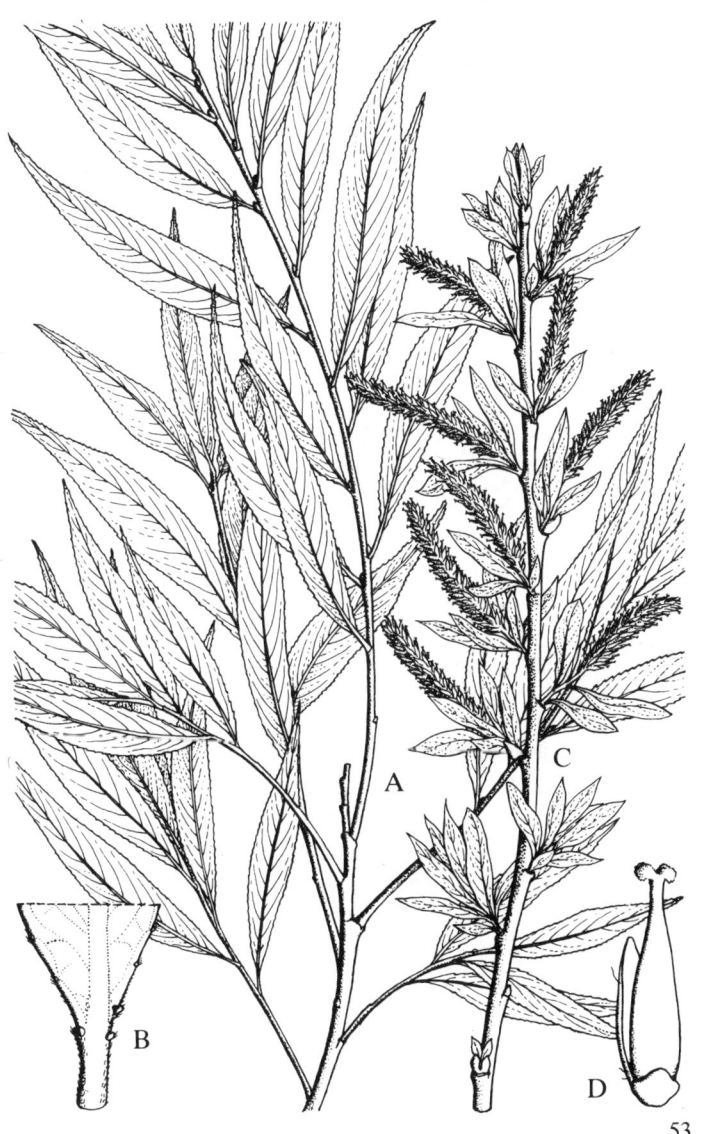

base cuneate, apex very slender and tapering, margins finely and rather regularly serrulate; lamina at first thinly pubescent on both sides, soon becoming glabrous, bright green above, persistently glaucous below, with rather obscure, ascending nervation; petiole usually less than 8 mm long, rather stout, generally with a few small glands near its apex; stipules usually small or wanting, caducous, narrowly ovate-acuminate, margins glandular-serrulate, and with internal (upper) surface usually sparsely glandular. Catkins appearing with the leaves in April, terminal on very short, spreading, leafy, lateral shoots, peduncle and rhachis softly villose. Catkins male, female or commonly androgynous, narrowly caudate-cylindrical, often somewhat curved, 3–4 cm long, 0.3–0.5 cm wide, dense-flowered; catkin-scales uniformly pale yellow, oblong-ovate, about 2 mm long, 1 mm wide, thinly hairy; stamens 2; filaments free, thinly hairy in the lower half; anthers shortly oblong, about 0.5 mm long, 0.4 mm wide, yellow; nectaries 2 in male flowers, the abaxial narrowly oblong or subcylindrical, the adaxial much broader, often emarginate; female flowers with a single, oblong, blunt or emarginate nectary; ovary subsessile or very shortly pedicellate, glabrous, shortly flask-shaped, about 2.5 mm long, 1 mm wide; style very short and indistinct; stigmas bifid, small, shortly oblong.

The commonest of the cultivated "Weeping Willows", nowhere truly naturalised, but so well known and widely planted that it must be mentioned. Virtually nothing is known of its history, save that Louis-Albert Dode described it in 1908 from cultivated material obtained from Messrs. Spaeth, the Berlin nurserymen, who had listed it from 1888 onwards in their catalogue, as *Salix vitellina pendula nova*. It is difficult to believe that the tree was not known before this date, but search through literature has failed to disclose an earlier reference.

True *Salix babylonica* L., from China, with very slender, glabrous branches, caudate-acuminate stipules, and short, subsessile catkins, is extremely rare, or possibly extinct in our area, having proved tender and short-lived in our climate. It has been replaced by a perplexing array of hybrids of which *S.* × *sepulcralis* nothovar. *chrysocoma* is, as already mentioned, quite the most popular. In addition to this *alba-babylonica* hybrid, one occasionally finds

---

A: leaves × 2/3; B: stipule × 8; C: ♂ catkins × 2/3; D: detail × 12; E: ♀ catkins × 2/3; F: detail × 12.

Salix × sepulcralis nothovar. **chrysocoma** 3a × 4

planted specimens of typical *Salix* × *sepulcralis* Simonk. (*S. babylonica salamonii* Carrière), presumed to be a cross between *S. alba* L. var. *alba* and *S. babylonica,* and thought to have originated in France about the middle of the last century. It is a sturdy tree, less pendulous than *S.* × *sepulcralis* nothovar. *chrysocoma,* with olive-brown twigs, and rather broader, less acuminate, duller green, glabrescent leaves. Although a handsome tree, *S.* × *sepulcralis* has never become popular, perhaps because it is less spectacular than *S.* × *sepulcralis* nothovar. *chrysocoma.*

4 × 2. Salix babylonica L. × S. fragilis L. = Salix × pendulina Wenderoth

A "weeping" tree, to about 12 m high, sometimes with the branches and twigs as slender and pendulous as in *S. babylonica,* sometimes (*S.* × *pendulina* var. *blanda*) distinctly less pendulous; bark deeply and coarsely fissured; twigs olive-brown, glabrous; buds narrowly ovate-acuminate, glabrous, glossy brown. Leaves lanceolate, 10–12 cm long, 1.5–2 cm wide, base broadly cuneate, apex slender and tapering, margins distinctly and often rather irregularly glandular-serrate; lamina glabrous or very soon glabrescent, dark lustrous green above, paler and rather glaucous below, with rather obscure, ascending nervation; petiole 0.8–1.5 cm long, glabrous or thinly pilose, generally with a few glands near its apex; stipules ovate-acuminate with glandular-serrate margins, caducous, the acumen very long and slender in *S.* × *pendulina* var. *blanda,* internal (upper) surface sparsely glandular. Catkins appearing with the leaves in April, terminal on short, leafy, lateral shoots; peduncle glabrous or subglabrous; rhachis rather densely pubescent. Catkins generally female, but often abnormal and androgynous, narrowly caudate-cylindrical, straight or slightly curved, about 3 cm long, 0.5 cm wide, dense-flowered; catkin-scales uniformly pale yellow, shortly oblong or oblong-lingulate, acute or obtuse, 2–3 mm long, 1 mm wide, thinly pilose; stamens (in androgynous catkins) apparently 2, with glabrous or pubescent-based filaments; nectaries 2 (or sometimes 1 in female flowers), the adaxial distinctly broader than the abaxial; ovary subsessile or distinctly pedicellate, glabrous or

---

A: leaves × 2/3; B: leaf base × 4; C: stipule × 3; D: ♀ catkins × 2/3; E: detail × 12; F: detail of ♂ and ♀ catkin × 3.

Salix × pendulina var. **elegantissima** 4 × 2

sparsely pubescent near the base, flask-shaped, about 2.5 mm long, 1 mm wide; style short but distinct, often 2-cleft above; stigmas oblong, shortly bifid.

At least three variants are found in gardens; all would appear to have originated in Germany (possibly at Marburg) early in the 19th century, but their origin is no less confused than their taxonomy and nomenclature. The commonest in Britain and Ireland, though much less frequent than *Salix* × *sepulcralis* nothovar. *chrysocoma*, is *S. elegantissima* C. Koch (here named *S.* × *pendulina* Wenderoth var. *elegantissima*), a tree with the extreme weeping habit of *S. babylonica*, but hardier and more satisfactory than *S. babylonica*, though less colourful than *S.* × *sepulcralis* nothovar. *chrysocoma*.

Typical *S.* × *pendulina* Wenderoth was first described in 1831 from material said to have been in cultivation in Marburg for twenty years. The author says it is very similar to *S. babylonica* in general appearance, but goes on to note that the leaves are entire and the ovaries adpressed-hairy, characters quite at variance with *S. babylonica*, and presumably influencing later authors in referring *S. pendulina* to synonymy under *S. purpurea* L. A specimen in the Kew herbarium, said to have been made from a living plant of *S. pendulina* sent by Wenderoth to Koch, is, however, quite clearly a *babylonica-fragilis* hybrid, differing from *S. pendulina* var. *elegantissima* only in having completely glabrous ovaries.

S. × *pendulina* Wenderoth var. *elegantissima* was first described (as *S. elegantissima*) by C. Koch in 1871, and was said to be widely distributed as a cultivated tree in N.E. Germany at this date. Koch thought it was a native of Japan, but modern Japanese authors regard it as a cultivated, non-indigenous tree, and one suspects it arose in Germany, probably at the same time as *S.* × *pendulina*. Koch also notes that the ovaries are quite glabrous (as in *S.* × *pendulina*), but von Seemen (1908) remarks that all the material he has seen has hairy ovaries, which, as is shown in our plate, would appear to be true also of our material of *S.* × *pendulina* var. *elegantissima*.

S. × *pendulina* Wenderoth var. *blanda* was described by N. J. Andersson in 1865 from specimens collected at Hanau in W. Germany. It is less "weeping" than *S.* × *pendulina* var. *elegantissima*, and also differs in having caudate-acuminate stipules and distinctly pedicellate ovaries. It is evidently not popular in European gardens, though it is to be seen in botanical collections at Kew and elsewhere. From descriptions it would appear to be the Wisconsin Weeping Wilow.

To assist identification, the following key summarizes the characters of *S. babylonica* and its hybrids:

1 Catkins subsessile, rarely exceeding 2 cm in length; ovaries shortly pyriform or ovoid, not tapering to apex
    4 S. babylonica
1 Catkins distinctly pedunculate, usually more than 2 cm long; ovaries flask-shaped, tapering to apex
  2 Leaves finely serrulate, pubescent or silky when young; ovaries shortly flask-shaped, not much longer than subtending catkin-scale
    3 Twigs golden or greenish-yellow, very pendulous
        3a × 4 S. × sepulcralis nothovar. **chrysocoma**
    3 Twigs brownish or olive, not very pendulous
        3 × 4 S. × sepulcralis var. **sepulcralis**
  2 Leaves distinctly and rather irregularly serrate; glabrous or at most very thinly hairy at first; ovary elongate flask-shaped, much longer than subtending catkin-scale
    4 Ovary shortly hairy near the base, subsessile; tree strongly "weeping"
        4 × 2 S. × pendulina var. **elegantissima**
    4 Ovary completely glabrous
      5 Ovary subsessile; stipules shortly acuminate; tree strongly "weeping"
        4 × 2 S. × pendulina var. **pendulina**
      5 Ovary distinctly pedicellate; stipules with a slender, caudate-acuminate apex; tree not strongly "weeping"
        4 × 2 S. × pendulina var. **blanda**

5. Salix triandra L.

Almond Willow

A robust, spreading shrub or small, bushy tree up to 10 m high; bark rather smooth, dark greyish, flaking off in large irregular patches (like the London Plane) to expose a reddish-brown underlayer; twigs glabrous, rather lustrous olive-brown, at first often conspicuously angled or ridged, becoming subterete with age, rather fragile at base; buds compressed dorsally, generally tapering to a rather slender apex, at first thinly and shortly pubescent, soon glabrous. Leaves lanceolate, oblong-lanceolate or narrowly elliptic, occasionally almost linear-lanceolate, 4–11(–15) cm long, 1–3(–4) cm wide, base rounded or cuneate, apex acute or long-acuminate, margins somewhat thickened, conspicuously and regularly serrate; lamina glabrous, dark rather dull green above, greenish or distinctly glaucous below, venation not very conspicuous; petiole usually less than 2 cm long, glandular at the apex, the glands sometimes enlarged and conspicuous; stipules often large and persistent, broadly auriculate, acute or shortly acuminate, 5–10 mm long, 3–5 mm wide, regularly or irregularly glandular-serrate, the upper (adaxial) surface often sprinkled with sessile glands. Catkins appearing with, or a little in advance of the leaves in April and May, and sporadically throughout the summer, erect on short lateral bracteate or leafy, lateral shoots; peduncle and rhachis shortly pubescent. Male catkins narrowly cylindrical, 2.5–5(–7) cm long, 0.3–1.2 cm wide, dense- or lax-flowered, showy, fragrant; catkin-scales shortly oblong or tongue-shaped, 1.5–2.5 mm long, 1–1.5 mm wide, concave, obtuse, uniformly pale yellow, thinly pilose or pubescent especially towards base; stamens 3, filaments 4–6 mm long, much exceeding the catkin-scales, pubescent towards base; anthers roundish or shortly oblong, less than 0.5 mm diam., bright yellow; nectaries 2, the inner shorter and thicker than the outer, oblong, truncate. Female catkins usually rather shorter and denser than the male, with similar catkin-scales; ovaries glabrous, rather broadly flask-shaped, about 2.5 mm long, 1.4 mm wide, distinctly pedicellate, with pedicels 1–1.5 mm long, sometimes lengthening to 3 mm or more in fruit; styles very short and inconspicuous; stigmas

---

A: leaves × 2/3; B: ♂ catkins × 2/3; C: detail × 8; D: ♀ catkins × 2/3; E: detail × 8.

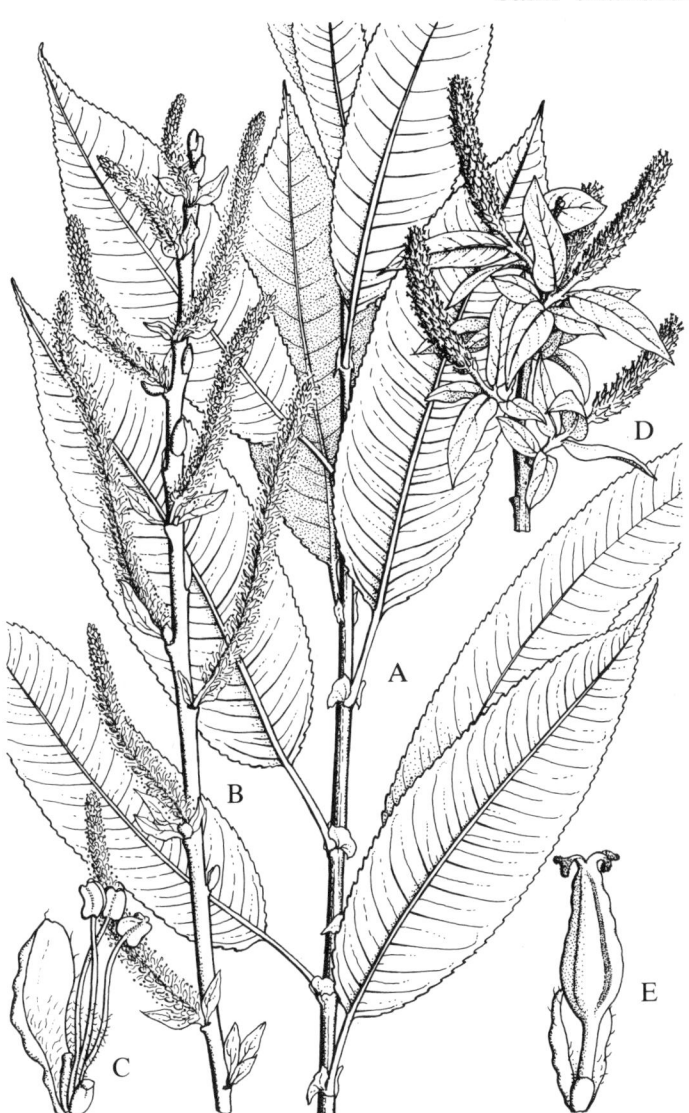

Salix triandra

short, spreading, emarginate or bifid; nectary 1, oblong, truncate, shorter than the pedicel.

Though locally abundant in wet ground south and east of a line connecting the Humber and Severn estuaries, and fairly frequent throughout England, the standing of *Salix triandra* must be regarded as questionable. It is one of the most important of the basket-makers' Willows, with a multiplicity of named varieties ('Black Maul', 'Grizette', 'Yellow Dutch', 'Mottled Spaniards', 'Sarda'—to name but a few) in cultivation, and a similar range of variants in the wild, many (or perhaps all) of them descendants of the Osier-growers' cultivars, often more or less obviously planted or relics of cultivation. The species is certainly not native in Ireland, and has little claim to be regarded as indigenous in Scotland or Wales.

In view of the very large numbers of variants in cultivation, it is not surprising that salicologists should have had difficulty in making satisfactory infraspecific classifications. Andersson (*1*) recognised six major variants, based on the colour of the upper and lower surfaces of the leaves, the length, breadth and density of the catkins and the habit of growth; some of these variants he in turn subdivided into broad-leaved and narrow-leaved forms, but the whole classification, though workable, is rather obviously mechanical and artificial, and few of the infraspecific taxa are maintained nowadays. Smith's attempt (*27*) to separate *S. triandra* L. and *S. amygdalina* L. at species level, chiefly on leaf characters, is even less successful and has been abandoned. Though it is tempting to make some sort of analysis of such a polymorphic species, the exercise is not a profitable one. *S. triandra* is just as variable elsewhere in Europe, where it is widespread; it also extends eastwards through Turkey and Iran to Central Asia and has close allies in China and Japan. It is one of the most attractive and fragrant of Willows, and deserves wider recognition as a garden shrub. The twigs have a pleasant flavour of rose-water when chewed.

5a. **Salix triandra** L. var. **hoffmanniana** Bab.

A much-branched bush, usually less than 4 m high; bark flaking; twigs very slender, glabrous. Leaves uniformly smaller than in typical *S. triandra,* 2–6(–7) cm long, 1–1.5(–2.5) cm wide, lanceolate, dark rather lustrous green above, paler, but not glaucous, below; petiole short, generally less than 1 cm long; stipules conspicuous, persistent, broad and often very blunt, sometimes almost orbicular. Catkins (usually male) smaller than in typical *S. triandra,* seldom more than 7 cm long and 0.7 cm wide.

The male plant is locally common in Dorset, Sussex and Surrey, and occurs here and there, often obviously planted, in other parts of our area, as far north as S. Uist. The female plant is much rarer, and records for it are often erroneous, but I have seen what appear to be correctly identified specimens in Leefe's herbarium (unlocalized, Kew) and from Oxted, Surrey (Fraser's herbarium, Kew). The female specimen mentioned by Linton "from a brookside not far from Shanklin Church, Isle of Wight [May 2] 1840" (Bromfield herbarium, Kew) is not var. *hoffmanniana,* but some other concolorous-leaved variant of *Salix triandra.* Genuine material of the variety is easily recognized in the field by its low twiggy growths and small leaves; in the herbarium the broad persistent stipules are distinctive, but it is not always easy to separate specimens of var. *hoffmanniana* from depauperate, concolorous-leaved forms of the typical plant.

Continental authorities ignore var. *hoffmanniana,* nor have I any satisfactory evidence of its occurrence outside our area. *S. triandra* var. *hoffmanniana* is, one suspects, little more than a distinctive cultivar, though, according to several authorities, it is rated poorly by Osier growers because of its very intricate mode of growth.

5 × 9. Salix triandra L. × S. viminalis L. = Salix × mollissima Hoffm. ex Elwert

This hybrid includes at least three varieties, two of which occur in our area; the three may be distinguished as follows:

1 Leaves entire or subentire
  2 Leaves linear-lanceolate; glabrescent; ovaries thinly hairy; styles about as long as stigmas   var. **hippophaifolia**
  2 Leaves lanceolate; pubescent at least below; ovaries densely hairy; styles much longer than stigmas   var. **mollissima**
1 Leaves serrate, soon glabrous and lustrous above; ovaries glabrous or very sparsely pubescent   var. **undulata**

What may be called the type variety, *Salix* × *mollissima* Ehrh. var. *mollissima*, is not known to grow in Britain, though one specimen collected by Miss E. Vachell, on May 17, 1904 (Fraser's herbarium, Kew) resembles it in having distinctly pubescent leaves, hairy ovaries and long styles. Unfortunately the specimen lacks mature foliage, but a further search in the neighbourhood of "Ely Bridge, banks of R. Ely, Glamorgan" might be rewarding. Otherwise, var. *mollissima* is widely, but locally, distributed in France, Germany, Denmark and Sweden, and possibly further eastwards into central Europe.

A: leaves × 2/3; B: ♂ catkins × 2/3; C: detail × 12; D: ♀ catkins × 2/3; E: detail × 12.

Salix triandra var. hoffmanniana

5 × 9 (a). Salix × mollissima Hoffm. ex Elwert var. hippophaifolia (Thuill.) Wimm.

A robust shrub 3–5 m high, with subglabrous or glabrescent twigs and acuminate, shortly puberulous buds; the leaves resemble those of *S. viminalis* in size and shape, being linear, or linear-lanceolate, up to 13 cm long and 1.5 cm wide, with a long acuminate apex and entire or very minutely glandular-denticulate margins; unlike the leaves of *S. viminalis,* however, they are soon glabrous or almost glabrous, dark shining green above, duller and paler below. Male and female catkins, produced with the leaves in April and May, seem equally common. The males are subsessile or shortly pedunculate, to about 3.5 cm long and 1 cm wide, with densely hairy, yellowish scales, and 2–3 stamens with yellow anthers. Not uncommonly a few female flowers are to be found intermixed with the males. The female catkins are more slender, rarely more than 5 mm wide, distinctly pedunculate with leafy bracts; the small, flask-shaped ovaries are at first shortly grey-pubescent, but the pubescence thins with age, and may become very sparse in over-mature examples; the styles are rather short, about as long as the 2-cleft stigmas.

A local plant, found here and there south of Notts. and Derby, but nowhere common, and not satisfactorily recorded from Scotland or Ireland. Leaf indumentum apart, it leans more towards *Salix viminalis* than *S. triandra,* but will seldom give any difficulty to the identifier.

Linton (*16*) tries to distinguish *S. trevirani* Spreng. by its broader, more glabrous leaves, longer catkins and glabrous ovaries. I cannot, however, follow the distinction, at least as regards British specimens; some so labelled are evidently *S. × mollissima* var. *hippophaifolia,* others are inseparable from *S. × mollissima* var. *undulata.*

A: leaves × 2/3; B: stipule × 4; C: ♂ catkins × 2/3; D: detail × 12; E: ♀ and mixed catkins × 2/3; F: detail of ♀ catkin × 2; G: mixed catkin × 2.

Salix × mollissima var. **hippophaifolia** 5 × 9

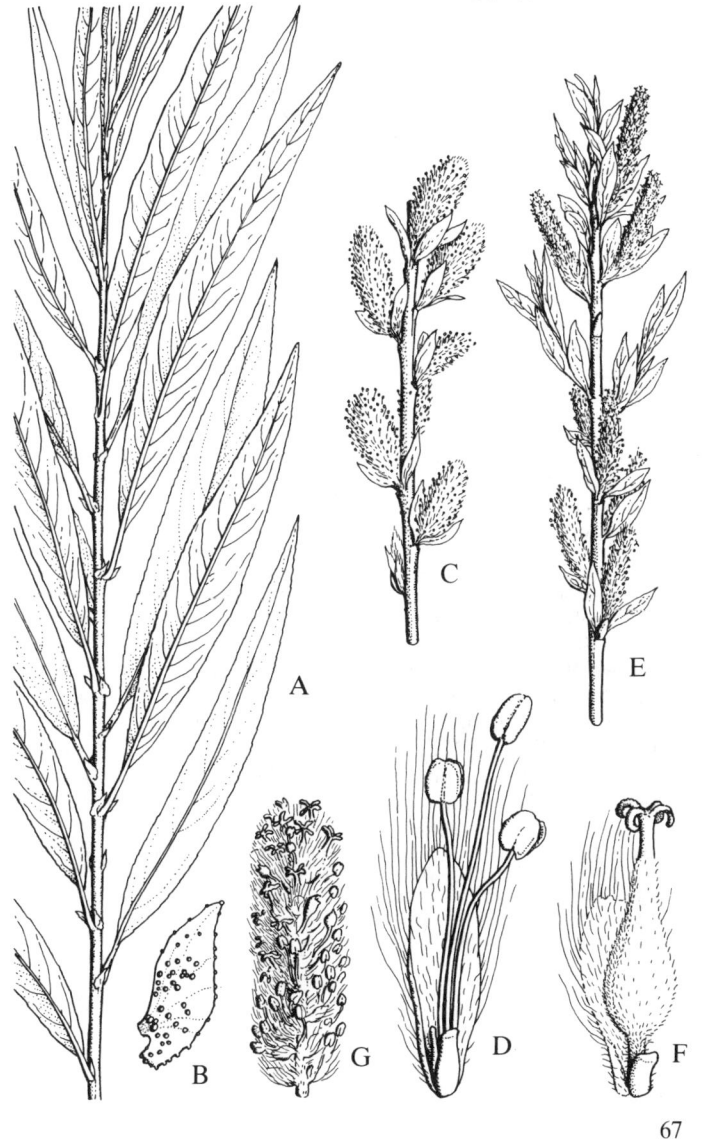

5 × 9 (b). Salix × mollissima Hoffm. ex Elwert var. undulata (Ehrh.) Wimm.

A tall shrub, commonly 4–5 m high, much resembling *S. triandra* in general appearance, with olive-brown or reddish, glabrous or glabrescent twigs and similar, flaking bark; the stipules are often conspicuous and persistent, thinly gland-dotted, with a drawn-out, acuminate apex and finely serrate margins; the leaves also resemble those of *S. triandra,* being dark green and rather lustrous above, paler or somewhat glaucous below, with distinctly, and often rather coarsely, serrate margins; the leaf-apex is, however, much more slender and acuminate than in *S. triandra*, so that the leaf as a whole appears longer and more slender, being frequently 10–12 cm long, but rarely much more than 1.5 cm wide. I have not seen male catkins, but the females (appearing with the leaves in April and May) are narrowly cylindrical, 3–4 cm long and about 0.5 cm wide, with pale catkin-scales rather sparsely clothed with long hairs, and relatively short, glabrous or subglabrous, very shortly pedicelled ovaries; the style is short, and about equal in length to the stigmas.

A common hybrid in southern England, but always an evident introduction or escape, and presumably a relic of Osier-holts, for it is a valuable basket-willow; it has also been recorded from a few vice-counties in Scotland and Ireland, and may prove more widespread in our area than present records indicate. It is easily confused with *Salix triandra,* but, without catkins, can be recognized by its long-acuminate leaves.

Linton (*16*) regarded *S.* × *undulata* as a hybrid between *S. alba* and *S. triandra,* but his reasoning is obscure, nor is it true to describe this hybrid as an "arborescent shrub", since it is normally a twiggy bush, though, like most lowland willows, it can occasionally develop a modest trunk.

The leaf-blade is commonly subtended by two small stipule-like auricles, as shown in the illustration.

A: leaves × 2/3; B: leaf base × 3; C: stipule × 4; D: ♀ catkins × 2/3; E: detail × 8.

Salix × mollissima var. undulata 5 × 9

6. Salix purpurea L.

Purple Willow

A shrub of variable habit, sometimes low, spreading and scarcely more than 1.5 m high, at other times more robust, occasionally attaining 5 m and forming a rounded bush or small tree; bark greyish, smooth, yellowish internally, and with very bitter taste; twigs glabrous, terete, slender, tough and flexible, usually yellowish or greyish, sometimes tinged red or purple; buds oblong-ovoid, rather elongate, acute, yellowish or reddish, sometimes a little pruinose. Leaves often opposite or subopposite, very variable in length and breadth, 2–8(–10) cm long, 0.5–3 cm wide, linear-oblong, oblanceolate or narrowly obovate, glabrous, or at first clothed with a thin, deciduous tomentum, rather dark green and dull or sublustrous above, paler or glaucescent below, becoming black on drying especially if collected when immature, apex acute or subacute, base gradually or rather abruptly cuneate, margins subentire or shortly serrate in the upper half; petiole usually very short; stipules small, narrowly oblong-acute, caducous and generally wanting even in the young growths. Catkins appearing before the leaves in March or April (or as late as May in northern counties), lateral, often opposite, sessile, erect or suberect, often curved, narrowly cylindrical, 1.5–3 cm long, 0.3–0.7 cm wide, dense-flowered, sometimes ebracteate, but usually with 2–3 small, narrow, acuminate, leaf-like bracts; rhachis densely pilose; catkin-scales very small, broadly and bluntly ovate or suborbicular, 1–1.5 mm long and almost as wide, thinly or densely pilose, blackish except at the base; nectary 1, shortly oblong, truncate or slightly emarginate. Male flowers with united filaments and anthers; filaments 2–4 mm long, much exceeding the catkin-scale, glabrous; anthers reddish or purple, shortly oblong, usually not much exceeding 0.5 mm in length. Female flowers with a small, sessile, hairy ovary about 1–1.5 mm diam., style very short and indistinct, stigmas shortly, and often bluntly, ovate, spreading, occasionally bifid. Capsule broadly ovoid, up to 4 mm long and 2.5 mm wide.

A locally common shrub of river-margins and wet ground, often planted, but with a much better claim to indigenous status than either *Salix triandra* or *S. viminalis*. It is widely distributed in

A: leaves × 2/3; B: ♂ catkins × 2/3; C: detail × 12; D: ♀ catkins × 2/3; E: detail × 12.

Salix purpurea 6

Britain and Ireland, and can be seen, far from habitations or cultivation, by streams or on damp hillsides.

Smith and Borrer attempted to distinguish four segregates: *S. purpurea* L., *S. helix* L., *S. lambertiana* Sm. and *S. woolgariana* Borrer; a fifth, *S. forbyana* Sm. is now considered a hybrid, *S. cinerea* × *S. purpurea* × *S. viminalis* (see p. 78). The others can be divided into two groups: *S. purpurea* and *S. helix* with narrow, linear-oblong or oblanceolate leaves, generally with entire or subentire margins; *S. lambertiana* and *S. woolgariana* with broader, oblong-obovate, often distinctly serrate, leaves. The last two (*S. lambertiana* and *S. woolgariana*) should certainly be united, since they cannot be distinguished by a single reliable character. *S. purpurea* and *S. helix* are distinguishable by habit and catkin-size, *S. purpurea* having slender, spreading or decumbent branches and small catkins seldom more than 20 mm long and 3.5 mm wide; *S. helix* is normally erect and more robust, with catkins usually more than 25 mm long and 5 mm wide. The stigmas of the (usually female) *S. purpurea* are short and ovate, those of *S. helix* (which is more commonly male) are said to be linear and divided. Unfortunately the female catkin selected by Smith to illustrate *S. helix* in *English Botany* (19: t. 1343) appears to have been taken from the hybrid *S.* × *forbyana* Sm., so that the stigma character, which sounds decisive, is in fact misleading. Furthermore the precise identity of the Linnean *S. helix* is difficult to establish, and scarcely tallies with Smith's interpretation. Modern authorities are prepared to admit that a recognizable distinction can be drawn between the narrow-leaved *S. purpurea* L. ssp. *purpurea* and the broad-leaved *S. purpurea* L. ssp. *lambertiana* (Sm.) A. Neumann ex Rechinger f., the latter originally collected, about the beginning of the 19th century, by A. B. Lambert on the banks of the R. Wylye at Boyton, Wiltshire. The distinction may be more striking on the Continent, where very slender, narrow-leaved variants of *S. purpurea*, often labelled *S. purpurea* L. var. *gracilis* Gren. et Godr., are common, and apparently indistinguishable from Smith's interpretation of *S. purpurea* L. In this country the prevailing variant is closer to that described by Smith as *S. helix,* which is, however, but one of a series of forms (or more probably cultivars) connecting narrow-leaved and broad-leaved plants.

Salix purpurea has an extensive distribution in Europe, though absent from Scandinavia and much of the U.S.S.R.; it has several very close allies in Asia.

A: leaves × 2/3; B: ♀ catkins × 2/3; C: detail × 12; D: leaves × 2/3; E: ♂ catkins × 2/3; F: detail × 12.

Salix purpurea (variants) 6

6 × 9. Salix purpurea L. × S. viminalis L. = Salix × rubra Huds.

A tall shrub or small tree, usually 3–6 m high, sometimes more, with greyish, fissured bark and loose, spreading branches; twigs tough, flexible, glabrous and rather lustrous yellowish-brown except when very young. Leaves linear or narrowly lanceolate, 4–12(–15) cm long, 0.8–1(–1.5) cm wide, at first closely tomentose-pubescent, but soon glabrous and dark, rather lustrous, green above, paler and subglabrous or persistently puberulous, but not sericeous, below; apex usually tapering to a slender acumen, base tapering or narrowly cuneate; margins narrowly recurved, flat or undulate, often distinctly but remotely serrulate, especially towards apex; stipules linear or subulate, minutely glandular-dentate, caducous and generally wanting except in the strongest and youngest growths. Catkins appearing before the leaves in late March and April, often rather crowded towards the tips of the branches as in *S. viminalis*, sessile, shortly cylindrical, about 2–3.5 cm long, 0.7–1 cm wide; catkin-scales usually short, blunt, black and hairy as in *S. purpurea*, rarely brownish as in *S. viminalis;* stamens with 2 free or partly united, glabrous, or sometimes thinly hairy, filaments; anthers red or yellow; ovary broadly ovoid or subglobose, equalling or slightly exceeding the subtending scale, densely grey-hairy; style short but distinct; stigmas 2, linear, undivided, erect or spreading, about twice as long as the style.

One of the commoner hybrids, found here and there throughout Britain and Ireland, but often a relic of Osier-beds, since it was formerly a popular Willow with basket-makers, and represented in cultivation by a wide range of cultivars. It varies within a narrow compass, however, and rarely gives much difficulty to the identifier, since the green undersurface of the leaves distinguishes it from *Salix viminalis*, while the pubescence (especially noticeable in young stems and leaves) rules out *S. purpurea*. *S. × rubra* is found almost throughout Europe.

A: leaves × 2/3; B: stipule × 4; C: ♂ catkins × 2/3; D: detail × 12; E: ♀ catkins × 2/3; F: detail × 12.

Salix × rubra 6 × 9

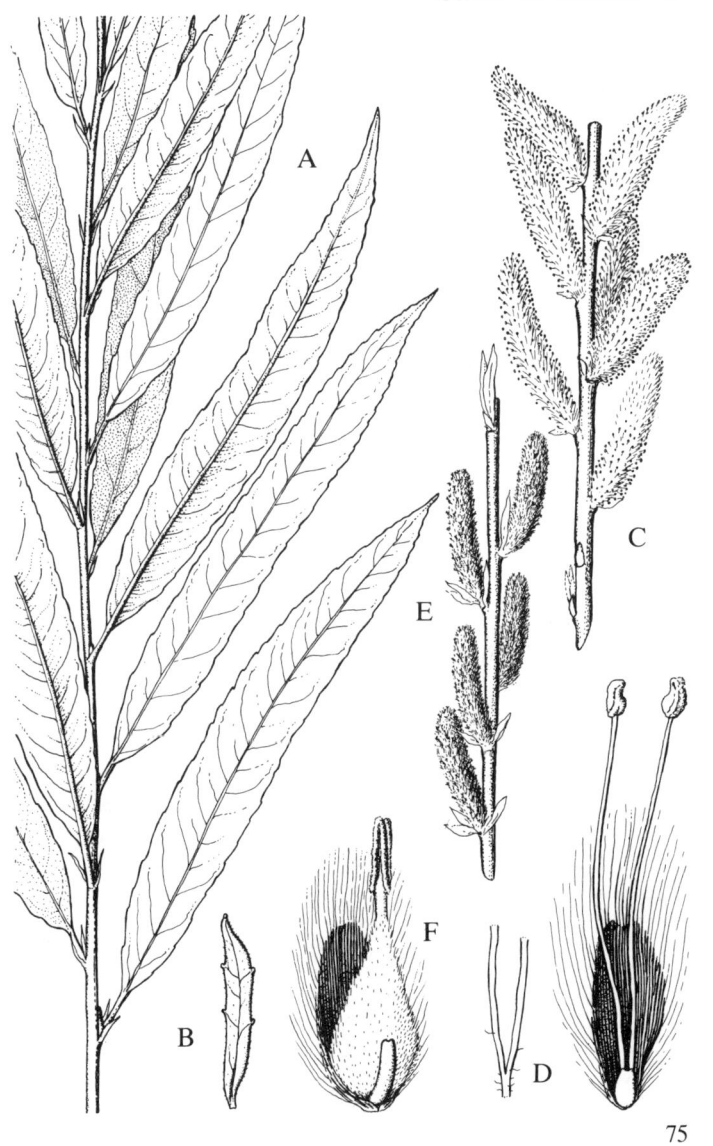

6 × 17. Salix purpurea L. × S. repens L. = Salix × doniana Sm.

A low spreading bush, 1–1.5 m high, with much-branched, glabrescent, rather lustrous, reddish-brown twigs; the leaves are, at first, very closely sericeous below, but soon become glabrous, with the upper surface bright, shining green, the lower distinctly glaucous. As in *S. purpurea,* some of the leaves tend to be opposite or sub-opposite, and, in common with both parents, the leaves generally blacken on being dried. The leaf-blade is oblanceolate, 2–4 cm long, 0.5–1.3 cm wide, with an abruptly acute or sometimes almost cuspidate apex and subentire or remotely serrulate, narrowly recurved margins. The petioles are very short or almost wanting, and stipules are rarely to be seen, even in the youngest growths. The catkins appear in advance of the leaves in March and April, and are sessile or very shortly pedunculate, with a few basal, sericeous or glabrescent, bracts; the males are shortly cylindrical, about 1–1.5(–2) cm long, 0.6 cm wide, with short, rounded hairy catkin-scales, which are yellowish at the base, reddish above and black-tipped; the 2 stamens may be free, or partly joined, or united to the apex, and the anthers are dark purple; the female catkins are very neat and regular, usually less than 1.5 cm long and 0.5 cm wide, with densely hairy, dark-tipped scales and short, subsessile or shortly stalked, densely white-tomentose ovaries, about 2 mm long; the styles and stigmas are both very short.

A: leaves × 2/3; B: ♂ catkins × 2/3; C: detail × 12; D: ♀ catkins × 2/3; E: detail × 12.

Salix × doniana (typical) 6 × 17

A very uncommon hybrid, first described by Smith (1828) but without any detailed localization, save that it was "sent from Scotland, as British, by the late Mr George Don, to the late Mr George Anderson. *Mr Borrer.*" Twenty years later William Gardiner recorded it from "Baldovan Woods", near Dundee (*Flora of Forfarshire*, 105) and, if correct, may have unwittingly solved the mystery of the origin of *Salix* × *doniana*. F. B. White says that by 1888 the majority of botanists had come to the conclusion that it was not a British plant, but, in the summer of that year, he found "undoubtedly wild specimens on the banks of the river Tummel, near Pitlochry, in Perthshire", and was able to restore the hybrid, with certainty, to the Scottish flora. Since then it has been found on dune slacks at Freshfield and Formby, South Lancs., where it was first seen, in 1947, by Miss P. A. Jones. The Lancashire specimens (illustrated here), though closely resembling typical *S.* × *doniana*, are more robust, with broader, less tapering leaves. They are possibly hybrids between *S. purpurea* and *S. repens* var. *argentea*.

?13 × 6 × 9. ?Salix cinerea L. × S. purpurea L. × S. viminalis L. = Salix × forbyana Sm.

An erect, vigorous shrub about 3–5 m high; shoots at first sparsely puberulous, soon becoming glabrous; year-old twigs yellowish, rather glossy; bark on main branches almost smooth, greyish. Leaves narrowly oblong-lanceolate or oblanceolate, 3–12 cm long, 0.8–2.5 cm wide, at first clothed with a thin, deciduous, whitish or reddish tomentum, but soon glabrous, dark lustrous green above, duller and slightly paler below, blackening when dried, apex acute, base cuneate, margins irregularly and obscurely serrulate; nervation ascending, rather obscure; petiole usually less than 8 mm long; stipules seldom present, narrowly linear-acuminate, with recurved, serrulate margins. Catkins appearing before the leaves in March and early April, sessile or subsessile, cylindrical, 1–3.5 cm long, 0.6–0.7 cm wide, suberect, usually rather crowded towards the tips of the shoots; bracts lanceolate, about 1 cm long, 0.3 cm wide, glabrous above, at first adpressed-sericeous below, but glabrescent;

A: leaves × 2/3; B: ♂ catkins × 2/3; C: detail × 12; D: ♀ catkins × 2/3; E: detail × 12.

Salix × doniana (Lancs. form) **6 × 17**

catkin-scales bluntly oblong or ovate-oblong, about 2 mm long, 1.5 mm wide, fuscous, rather densely clothed with white silky hairs; nectary single, about 0.5 mm long, narrowly oblong, apex slightly emarginate; female flowers with a subsessile, shortly pyriform, densely white-pilose ovary about as long and as wide as the catkin-scale; style distinct, about 0.4 mm long; stigmas narrowly oblong, about as long as the style; male flowers with 2 free or partly united filaments about 4 mm long and shortly pubescent towards the base; anthers oblong, about 0.3 mm long, 0.2 mm wide, yellow, apparently tinged reddish or crimson towards the apex. Fruit not seen.

This distinctive hybrid takes the name from the Rev Joseph Forby, who collected specimens at Fincham, Norfolk, sometime about the beginning of the 19th century, and sent them to J. E. Smith. It was probably well known to Willow-growers long before this date, for Smith notes "highly valuable, as an osier, for the finer kinds of basket work". It is now known to occur, here and there, over much of England, southern Scotland and in north-western Ireland. Until 1959, I had only seen female plants, but in that year male specimens were sent to me by Mr R. C. L. Howitt, from material found at Newark, Notts. The male is, however, evidently very rare indeed. *Salix × forbyana* is not well known on the Continent, though I have seen specimens from France, Germany, Austria and Switzerland. In all instances I suspect, it is either cultivated, or a relic of cultivation. At first sight the glossy leaves may be mistaken for those of *S. triandra,* but the general absence of stipules, and the obscure and irregular leaf-serration, usually serve to rule out this identification, and the catkins are alone sufficient to establish the close affinity with *S. purpurea*. The crowded catkins and rather elongate stigmas indicate *S. viminalis,* and *S. × forbyana* may possibly be nothing more than a variant of the well-known *S. × rubra* Huds. There is, however, something distinctive in the leaf shape, and in the indumentum of the youngest leaves, which may be attributable to the presence of *S. cinerea*.

A: leaves × 2/3; B: ♂ catkin × 2/3; C: detail × 12; D: ♀ catkin × 2/3; E: detail × 12.

Salix × forbyana 13? × 6 × 9

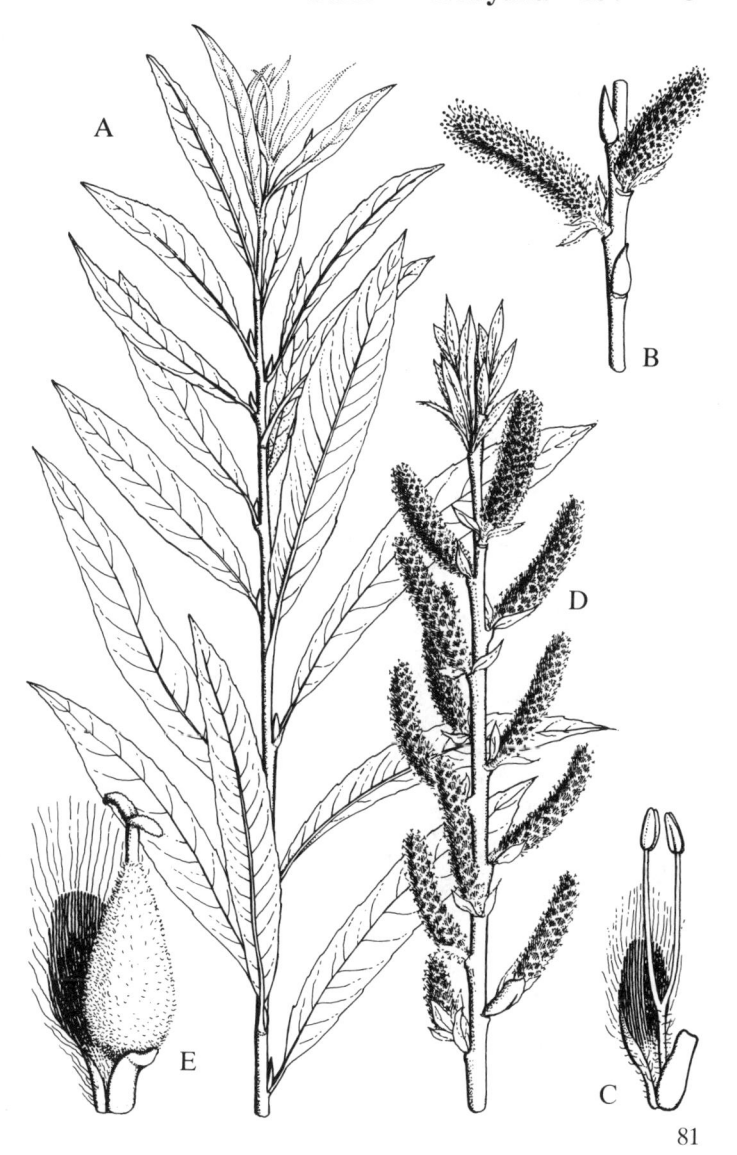

7. Salix daphnoides Vill.

A large shrub or small tree, usually 6–8 m high, occasionally up to 12 m, with a rounded crown, erect or spreading branches, and rather smooth, greyish bark; twigs glabrous, lustrous, dark reddish-brown, when young generally pruinose with a dense glaucous bloom; buds large, pointed, compressed dorsally, dark crimson, glabrous or more usually with the basal part covered with stiff, erect, deciduous hairs. Leaves oblong or narrowly obovate (4–)7–12(–14) cm long, (1–)2–3(–4) cm wide, at first thinly woolly, soon glabrous, dark lustrous green above, glaucous below, apex shortly acuminate, base broadly cuneate, margins distinctly and rather regularly glandular-serrate, narrowly thickened; petiole well developed, 0.7–2 cm long, rather robust, canaliculate and densely pilose above; stipules often persistent, narrowly ovate, long-acuminate, up to 12 mm long and 5 mm wide, margins distinctly glandular-serrate, internal surface usually covered with large sessile glands. Catkins appearing before the leaves in February and March, sessile, erect or suberect, often rather crowded, shortly cylindrical, dense, 2–4 cm long, 0.8–1.8 cm wide, ebracteate, or with a few, inconspicuous, densely pilose bracts; rhachis pilose; catkin-scales broadly ovate-elliptic, shortly acute, about 2 mm long, 1.8 mm wide, dark blackish-brown, densely hairy, the hairs of the male flowers often 7 mm long; nectary-scale 1, narrowly oblong, truncate or slightly emarginate. Male flower with 2 free stamens; filaments 7–8 mm long, glabrous, slightly exceeding the hairs of the catkin-scale; anthers narrowly oblong, about 0.8 mm long, 0.5 mm wide, yellow. Female catkins smaller and less decorative than the males, with shorter hairs on the catkin-scales; ovary very narrowly flask-shaped, shortly pedicellate, tapering to a slender neck, about 4 mm long, 1 mm wide, style up to 0.8 mm long, sometimes inconspicuous, stigmas narrowly oblong, about 0.6 mm long, erect or spreading, occasionally bifid. Capsule narrowly ovoid, about 4 mm long, 1.5 mm wide.

Although widely planted, *Salix daphnoides* is seldom seen far from houses or gardens, and cannot be considered a native or even a naturalized species, though it has been given a place in the British plant-list for a very long time. Though identifiable by its large,

A: leaves × 2/3; B: stipule × 6; C: ♂ catkins × 2/3; D: detail × 8; E: ♀ catkins × 2/3; F: detail × 8.

Salix daphnoides 7

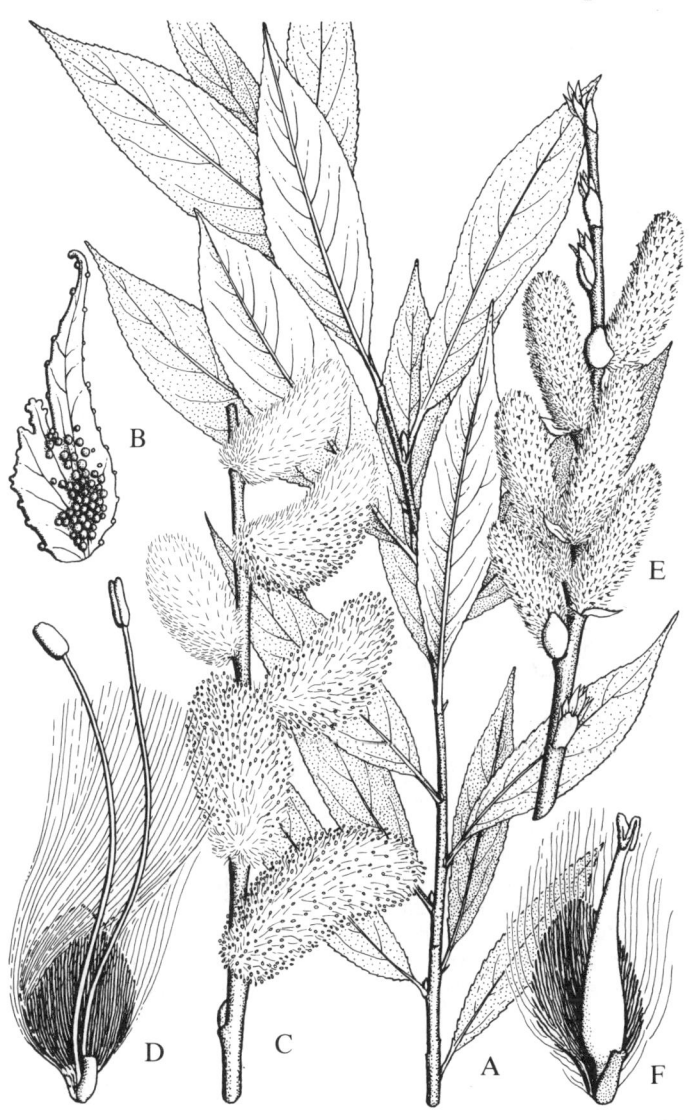

showy catkins and coriaceous, lustrous leaves, *S. daphnoides* is most readily distinguished from our other Willows (except *S. acutifolia* Willd.) by its pruinose twigs; even when the bulk of the "bloom" has been rubbed off, a certain amount will generally persist immediately below the buds. As with most willows, the male catkins are much more ornamented than the female, and males are, in consequence, more frequently cultivated than females.

S. daphnoides has a wide but discontinuous distribution in central Europe, from the Baltic states south to Piedmont, and from eastern France to the Balkans. The closely allied *S. daphnoides* var. *norvegica* Ag. is recorded from Norway and Sweden. It should be noted that this distribution, quoted by a number of European authors, is rejected in *Flora Europaea* (1: 54), which says, "Norway, Sweden; widely planted especially in C. & E. Europe, but rarely naturalized".

8. Salix acutifolia Willd.

Violet Willow

A low tree or large bush, seldom more than 10 m high, with a broad crown and spreading branches; twigs usually long, slender and rather pendulous, glabrous, pruinose and becoming violet-coloured in winter. Leaves drooping on long, slender petioles, linear-lanceolate, commonly 15 cm long, seldom more than 2 cm wide, quite glabrous, lustrous green above, glaucescent at first, but becoming greenish below, apex with a long, tapering acumen, base narrowly cuneate, margins conspicuously glandular-serrate. Petiole sometimes 3 cm long, slender, glabrous, canaliculate above; stipules conspicuous, lanceolate-acuminate, with serrate margins, persistent, often (but not always) glandular on the inner surface. Catkins appearing with the leaves, or usually a little in advance of them, in March or April, very similar to those of *S. daphnoides,* seldom exceeding 3.5 cm in length or 1.2 cm in width, sessile, with inconspicuous, hairy bracts; catkin-scales blackish, narrowly ovate-elliptic, about 1.3 mm long, 1 mm wide, densely pilose with long silky hairs; nectary-scale linear or narrowly oblong, truncate. Male

---

A: leaves × 2/3; B: stipule × 4; C: ♂ catkins × 2/3; D: detail × 8.

Salix acutifolia 8

flowers with 2 stamens; filaments 7–8 mm long, glabrous, considerably exceeding the hairs of the catkin-scales, anthers oblong, about 0.6 mm long, 0.5 mm wide, golden-yellow. Female plants do not seem to occur in Britain; to judge from continental specimens, the catkins are very similar to those of *S. daphnoides,* but usually smaller and more slender with hairy catkin-scales and narrowly flask-shaped, tapering, glabrous ovaries.

So closely allied to *Salix daphnoides* Vill. that some authors prefer to regard it as a subspecies (ssp. *acutifolia* (Willd.) Dahl) or as a variety (var. *acutifolia* (Willd.) Doell) of this species; it is, at least in Britain, readily distinguished by its habit and foliage.

According to Andrews (*Botanist's Repository,* 9: t. 581 (1809)), *S. acutifolia,* then known as *S. violacea,* was introduced from Russia or Siberia by "Mr John Bell of Sion Gate about the year 1798". It is still found in gardens, but not commonly, and was formerly to be seen in some quantity, but evidently planted, by the side of the Thames near Mortlake and Chiswick. Specimens found on sand-dunes at Ainsdale, S. Lancs., were distributed by the Botanical Exchange Club in 1915, under the name *S. daphnoides* Vill. var. *pomeranica* (Willd.) Koch; this latter is evidently one of the small-leaved variants of *S. daphnoides* (*B.E.C. Rep.,* 4: 371 (1916)).

S. acutifolia is said to have a wide distribution in the U.S.S.R., ranging from the Baltic to Central Asia, with close relatives in E. Siberia and Mongolia.

9. Salix viminalis L.

Osier

A tall shrub or small tree, usually 3–6 m high, occasionally attaining 10 m; bark greyish-brown, fissured; branches erect or suberect, usually forming a rather narrow, truncate or rounded crown. Twigs often long and straight, very tough and flexible, at first densely ashy-pubescent, becoming smooth and rather lustrous yellowish-

---

A: leaves × 2/3; B: stipule × 4; C: ♂ catkins × 2/3; D: detail × 8; E: ♀ catkins × 2/3; F: detail × 8.

Salix viminalis

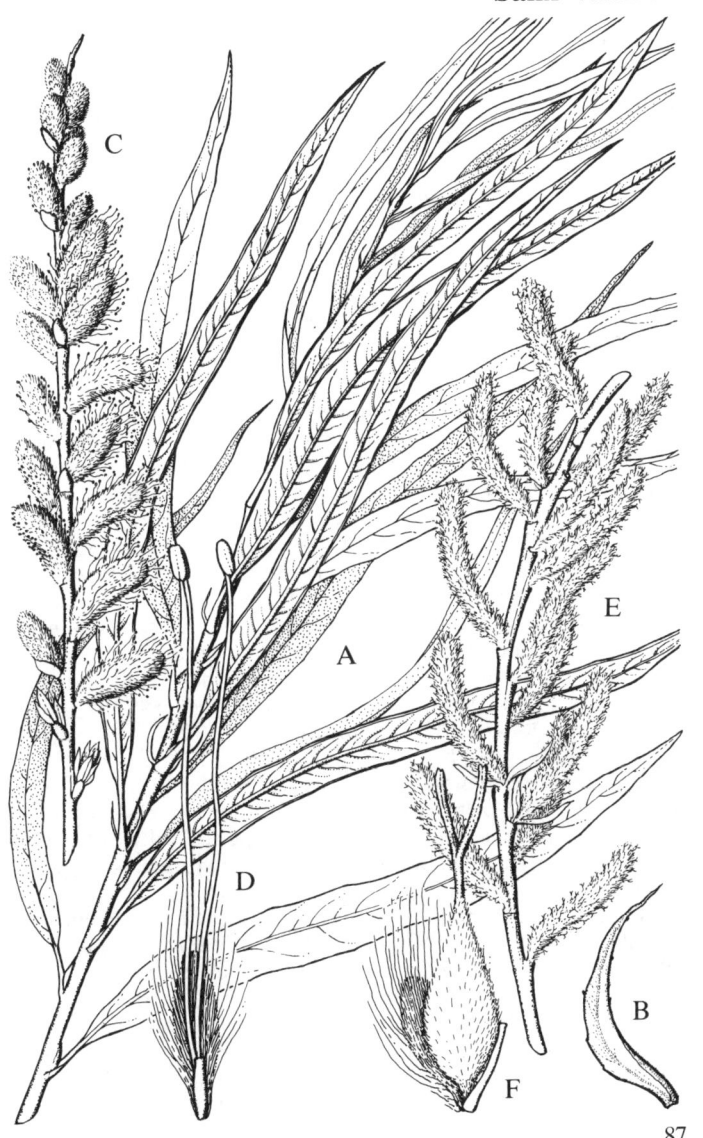

brown or olive-coloured; buds ovoid, obtuse or subacute, 3–7 mm long, 2–5 mm wide, at first densely and shortly pubescent, becoming glabrous or subglabrous and yellowish or reddish-brown. Leaves linear or narrowly linear-lanceolate, 10–15(–18) cm long, 0.5–1.5(–2.5) cm wide, dull green and thinly puberulous above, silvery below with a dense indumentum of short, adpressed silky hairs; apex tapering to a long slender acumen; base narrowly cuneate, margins appearing entire, narrowly recurved or revolute, often conspicuously undulate; petiole short, rarely exceeding 1 cm, pubescent, canaliculate above; stipules linear or narrowly lanceolate-acuminate, subentire or obscurely glandular-dentate, often falcate, up to 10 mm long, seldom more than 2 mm wide, caducous and often wanting on slender shoots. Catkins appearing before the leaves in late February, March or early April, sessile or subsessile and generally crowded towards the tips of the twigs, erect or somewhat recurved, narrowly ovoid or cylindrical, 1.5–3 cm long, 0.5–1 cm wide, dense-flowered, often apparently ebracteate, but with 2 or 3 small, densely sericeous, oblong-lanceolate bracts; rhachis densely pilose; catkin-scales narrowly ovate-elliptical, about 2 mm long, 1 mm wide, obtuse or subacute, densely pilose, mostly reddish-brown but sometimes blackish; nectary 1, narrowly oblong or linear-cylindrical, truncate. Male flowers with 2 free stamens; filaments glabrous, up to 1 cm long, much exceeding the catkin-scale; anthers oblong, yellow, about 0.5 mm long, 0.2 mm wide. Female flowers with a subsessile, flask-shaped, densely tomentose ovary, about 2 mm long and 1 mm wide; style distinct, about 0.8 mm long; stigmas linear, undivided or occasionally 2-cleft, spreading, up to 1.5 mm long. Capsule narrowly flask-shaped, tomentose, up to 6 mm long and 2 mm wide.

One of the commonest of our Willows, and one of the earliest to flower; the male plant, with its crowded, sub-terminal spikes of yellow catkins, can be recognised without difficulty, even at a distance, at a time when the landscape is otherwise almost devoid of colour.

Although so common everywhere, the status of *Salix viminalis* in Britain and Ireland is very questionable. It is always more or less obviously planted in Ireland and Scotland, and though, as Linton (*16*) remarks, it is "perhaps indigenous in eastern and central England", one rarely finds it except in an unnatural environment, and often as an obvious relic of former cultivation. It is seldom seen except at low altitudes. In western Europe *S. viminalis* is again mostly planted, or an escape or relic of cultivation, and it is not until one moves east to Russia that it begins to fit comfortably into natural communities; here (according to the *Flora of the U.S.S.R.*, 5: 133) it is found nearly all over the Soviet Union, from forest-tundra to the desert-steppe region, always along river-banks, where it forms dense and often extensive thickets. Since it is one of the favourite species with basket-makers, and since basket-making is one of the oldest of crafts, one is tempted to conjecture that it may have spread westwards with early man.

S. viminalis (perhaps because it is not strictly indigenous) is one of the least variable of our Willows. Some authors still maintain *S. viminalis* var. *linearifolia* Wimm. et Grab. (var. *angustissima* Coss. et Germ.), a depauperate state, with slender twigs and very narrow leaves, but almost any form of *S. viminalis* will sink into this condition if sufficiently neglected. *S. viminalis* var. *intricata* Leefe ex Bab. with a short style and "stigmas from the first cloven, reflexed and entangled" is more noteworthy, but hardly more important; var. *stipularis* Leefe ex Bab. has large, denticulate stipules and long stigmas, but is otherwise unremarkable. Of greater taxonomic significance is the readiness with which *S. viminalis* will hybridize with other Willows, providing, through its hybrid with *S. triandra,* a link between the Willows of the *S. fragilis* group and the Sallows.

10. Salix × calodendron Wimm.

A tall erect shrub or small tree, rarely more than 10 m high, resembling *S. viminalis* in habit, with similar erect branching and a rather narrow, truncate or blunt crown; twigs densely and persistently ashy-pubescent or tomentose, becoming fuscous and glabrescent only in their second year, the underlying wood marked with prominent longitudinal *striae* as in *S. cinerea* and *S. aurita*; buds bluntly ovoid, 7–8 mm long, 5–6 mm wide, densely pubescent, dull fuscous-brown. Leaves oblong-elliptic, usually about 10 cm long and 3 cm wide, occasionally up to 15 cm long and 5 cm wide, dull green and thinly pubescent above, ashy-grey below and rather more densely pubescent with longer hairs; midrib and lateral nerves rather prominent below; often reddish; apex acute or shortly acuminate; base rounded or broadly cuneate; margins narrowly recurved, obscurely and remotely glandular-serrulate; petiole stout and rather rigid, 1–1.5 cm long, densely pubescent and canaliculate above; stipules ear-shaped, conspicuous, 8–15 mm long, 5–7 mm wide, acuminate, often falcate, with recurved, glandular-denticulate margins. Catkins appearing before the leaves in March and early April, rather crowded towards the tips of the twigs as in *S. viminalis,* cylindrical, elongate, subsessile or commonly shortly stalked, 3.5–5(–7) cm long, 0.8–1 cm wide, erect or somewhat spreading; bracts foliaceous, villose, 10–12 mm long, 3–4 mm wide; peduncle and rhachis densely villose; catkin-scales broadly ovate, acute or obtuse, about 2 mm long, 1.5 mm wide, dark blackish-brown, densely hairy. Male flowers unknown. Female flowers with an ovoid or bluntly flask-shaped, densely white-tomentose, subsessile or shortly pedicellate ovary 3–4 mm long, 1.5 mm wide near base; style distinct, about 0.8 mm long; stigmas narrowly oblong or linear-oblong, about as long as the style, generally undivided. Capsule and seeds unknown.

I have previously dealt, at some length, with the confused nomenclature of this Willow (*Watsonia* 2: 243–245 (1952)) and with the various conjectures as to its origin. All known material is female, and all would appear to represent a single clone. It is apparently quite sterile, and remains so even when planted close to male bushes of *Salix viminalis* or *S. caprea*. Unless one is prepared to accept the mysterious *Salix dasyclados* Wimm. as a species, the most likely parentage for *S.* × *calodendron* is *S. caprea* × *S. cinerea* × *S. viminalis,* since it shares to some degree the characteristics of all three. An affinity with *Salix aegyptiaca* L., from western Asia, is also possible; this has markedly striate wood, and *caprea*-like leaves, with the nervation often reddish below as in *S.* × *calodendron*.

S. × *calodendron* is widely distributed in Britain, from Surrey to N. Aberdeen, and will probably also be found to occur over much of southern and eastern Ireland. It is seldom seen in any quantity, mostly growing as a solitary bush, and often as an apparent introduction, though why it should be planted is obscure, since it is neither particularly ornamental nor useful. It would seem to have been grown in Ireland as the "Jargomel Sallow" (*29* p. 359) and (see under *S. fragilis* var. *russelliana*) was supposed to have been introduced from Holland. Unfortunately Wade included more than one species in his interpretation of *S. acuminata* Sm. (= *S.* ×

A: leaves × 2/3; B: stripped twig × 2/3; C: catkins × 2/3; D: detail × 8.

Salix × calodendron

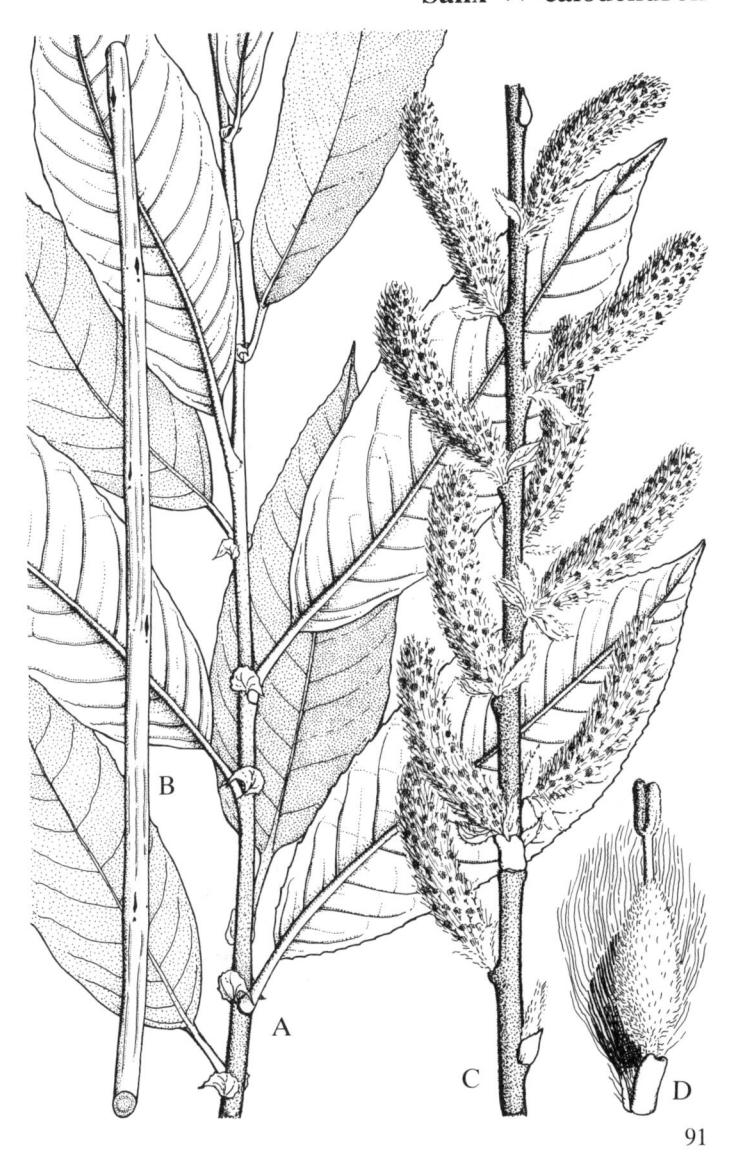

calodendron Wimm.) and the identity of the Jargomel (or Gorgomel) Sallow is debatable. The derivation of the name "Jargomel" is equally unclear.

There are few recent records for *S.* × *calodendron* from the Continent, but it is known, as a cultivated plant or an escape from cultivation, in Germany (Berlin), Poland (Wrocław), Sweden (Göteborg, Liban) and Denmark (Copenhagen).

11. Salix elaeagnos Scop.

An erect shrub up to 6 m high, or sometimes a slender, much-branched tree; growths of the current year densely whitish- or greyish-pubescent or tomentose, year-old twigs glabrous, yellowish-brown or reddish; buds small, subacute or obtuse, thinly pilose; leaves narrowly linear (narrower than those of any other Willow found in Britain), rather crowded, 5–15 cm long, generally less than 0.8 cm wide, subcoriaceous, dark shining green above, conspicuously white-tomentose below, tapering to the base, and to the finely acuminate apex; margins entire, narrowly recurved; nervation obscure; petiole very short and obscure, usually less than 5 mm long; stipules generally absent. Catkins appearing in April and early May, a little in advance of the leaves, generally numerous and rather crowded towards the tips of the twigs (as in *S. viminalis*), subsessile; bracts 2–4, foliaceous, linear, up to 20 mm long, rarely more than 2.5 mm wide, thinly tomentose on the lower surface; catkin-scales narrowly oblong, 3–4 mm long, 1 mm wide in female flowers, often much shorter and rounded in male flowers (from Central Europe), reddish or fuscous-tipped, subglabrous, or at most very sparsely pilose; male catkins spreading, up to 3 cm long, 0.5 cm wide, rather dense; stamens 2, long-exserted, filaments 5–7 mm long, united at base or to beyond the middle, thinly pilose towards base, anthers yellow, oblong, about 0.5 mm long, 0.3 mm wide; nectary shortly oblong, about 0.6 mm long, subtruncate. Female catkins usually rather shorter and more erect than the male, becoming rather lax with age, ovary narrowly flask-shaped, 4–5 mm long, 1–1.5 mm wide, glabrous, usually tapering gradually to an indistinct style, sessile or with a very short puberulous pedicel;

A: leaves × 2/3; B: detail × 4; C: ♂ catkins × 2/3; D: detail × 8; E: ♀ catkins × 2/3; F: detail × 8.

Salix elaeagnos

stigmas slender, about 0.6 mm long, generally cleft into 4 lobes. Capsule up to 5 mm long, 2 mm wide at maturity, remaining sessile or subsessile.

A native of central and southern Europe, western Asia and western North Africa, said by Loudon to have been introduced in 1821, and certainly growing at Woburn in 1829. Interestingly, the specimen figured as *Salix incana* by Forbes (*9* p. 179) is a male, while all the material recently collected in this country appears to be female. It is readily distinguished from all our other Willows save *S. viminalis* by its narrow tapering leaves, which are matt-white below (not sericeous). Female specimens are at once distinguished from female *S. viminalis*, with which it is occasionally confused, by the glabrous ovaries.

S. elaeagnos (presumably so named from a vague resemblance between its leaves and those of *Elaeagnus angustifolia*) is not infrequently cultivated, and occasionally turns up as a garden waif or reject; it can scarcely be considered an established member of our flora.

12. Salix caprea L.

Goat Willow

A tall shrub or small tree, rarely more than 10 m high; bark greyish-brown, irregularly fissured; branches spreading or ascending, forming a rather open crown; twigs robust, rigid, at first thinly pubescent, soon becoming glabrous and often yellowish or greyish-brown in their first year; underlying wood without *striae*; buds bluntly ovoid, up to 10 mm long and 8 mm wide, soon glabrous and glossy, those of the catkins exceptionally swollen and conspicuous, often yellow or stained with red. Leaves broadly oblong, obovate or elliptic, sometimes almost orbicular, 5–12 cm long, 2.5–8 cm wide, occasionally larger on coppiced shoots, dull green and thinly pubescent or subglabrous above, ashy-grey and softly tomentose or

---

A: leaves × 2/3; B: leaf × 2/3; C: stripped twig × 1; D: stipule × 3; E: ♂ catkins × 2/3; F: detail × 8; G: ♀ catkins × 2/3; H: detail × 8.

Salix caprea 12

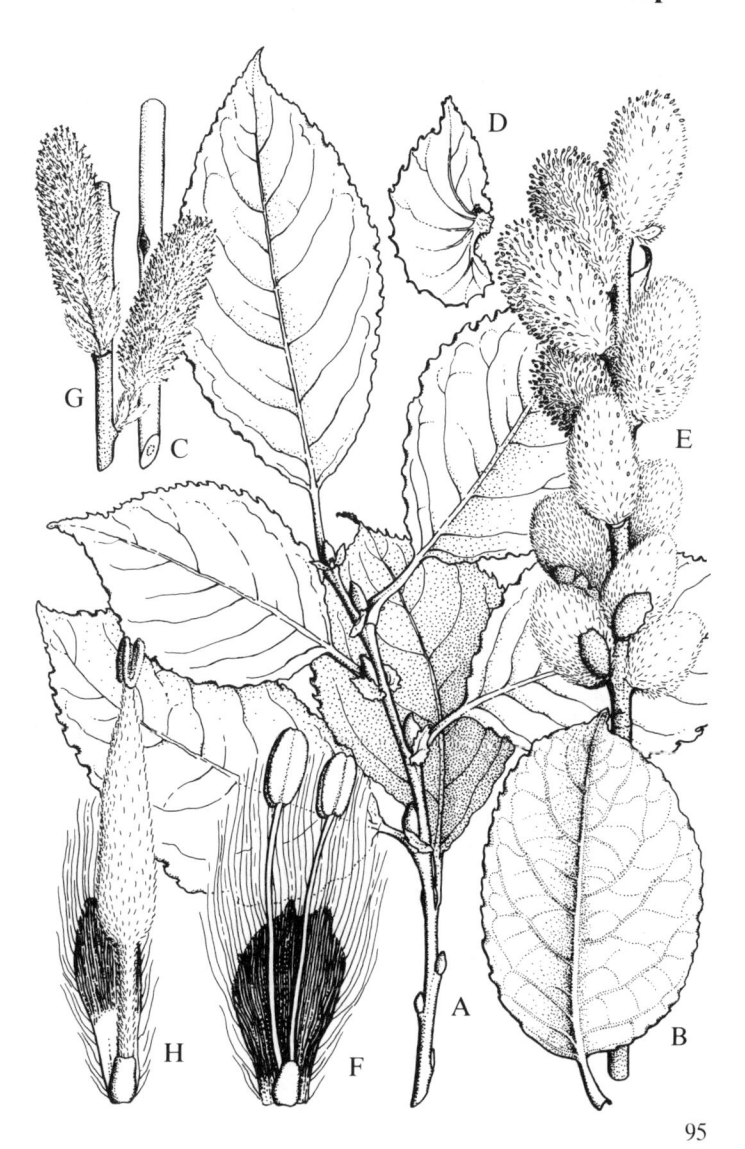

pubescent below, generally with prominently reticulate venation; apex shortly acute; base broadly cuneate, rounded or shallowly cordate; margins irregularly undulate-serrate, with short, gland-tipped teeth; petiole 0.8–2.5 cm long, rather rigid, pubescent, canaliculate above; stipules ear-shaped, 8–12 mm long, 3–8 mm wide, shortly acute, with undulate-serrate margins, sometimes persistent, but generally shed before the leaves are fully developed. Catkins appearing before the leaves in March and April, often crowded near the tips of the twigs, ovoid or shortly cylindrical, 1.5–2.5 cm long, 0.8–1.8 cm wide, erect; bracts small, inconspicuous, densely villose; catkin-scales broadly ovate, acute or obtuse, about 2 mm long, 1.8 mm wide, blackish, densely hairy. Male flowers with 2, long-exserted stamens; filaments up to 1 cm long, glabrous, anthers oblong, golden-yellow, about 1.3 mm long, 0.8 mm wide; nectary single, oblong, about 0.5 mm long, 0.3 mm wide. Female flowers with a narrow, tapering flask-shaped, pedicellate, densely silvery-sericeous ovary, 4–5 mm long, 1–1.5 mm wide, pedicel at first about 1.5 mm long, lengthening considerably in fruit; style wanting or short and indistinct, stigmas erect or suberect, about 0.5 mm long, oblong, sometimes shortly 2-cleft. Capsule up to 10 mm long, 2 mm wide, thinly sericeous.

The first to flower of the indigenous Willows, and a very conspicuous feature of the leafless, early spring landscape. The Goat Willow is generally easy to recognize at all seasons, by its swollen, colourful catkin-buds, by its handsome sessile catkins, or by its unusually broad, Apple-like leaves. It is not in itself a very variable species, and few of the forms described (mostly based on leaf-shape) are worth serious attention; identification of *Salix caprea* is, however, rendered rather more difficult by the frequency of hybrids, especially with *S. cinerea* L. ssp. *oleifolia* Macreight. It would appear that these hybrids can back-cross freely with either parent, and a complete range of specimens can often be collected showing all the transitional characteristics. The prevalence of this hybrid is no less troublesome in the field than in the herbarium, and it is often difficult to be quite certain that a particular specimen, outwardly answering to *S. caprea*, may not be partly tainted with *S. cinerea* ssp. *oleifolia*. The colour of the year-old twigs, and their relative sturdiness (those of pure *caprea* being noticeably thick and stocky) will often settle a doubt where leaves and catkins are an unsatisfactory guide. The winter twigs of *S. caprea*, greenish-yellow stained reddish-brown along the exposed surfaces, are quite distinct from those of other lowland willows.

The Goat Willow is locally common throughout Britain and Ireland, usually growing in hedgerows, or by woodland margins, or on rocky lake shores; it is more tolerant of dry situations than *S. cinerea* or its subspecies, and seems to show a slight preference for calcareous soils, though it would be an exaggeration to call it a calcicole. It was a common invader of bombed-sites in urban areas during and after World War II. Druce says it occurs at altitudes of 850 m (2800 ft), but it is mostly lowland, and one suspects that high altitude records may be more correctly referred to the following, *S. caprea* var. *sphacelata*.

Abroad, *S. caprea* is found almost throughout Europe, and eastwards through Turkey to Central Asia, with close allies in China and Japan. In Scandinavia and Russia, and at high altitudes in the central European Alps, *S. caprea* var. *sphacelata* is locally predominant.

2a. Salix caprea L. var. sphacelata (Sm.) Wahlenb.

Often a small, gnarled shrub, but occasionally becoming a small tree up to 10 m (or more) in height; twigs at first pubescent, soon becoming glabrous and dull fuscous-brown; buds as in *S. caprea* L. var. *caprea* but often darker, reddish, and rather persistently puberulous; leaves rather narrowly obovate-elliptic, sometimes broader, 3–7 cm long, 1.5–4.5 cm wide, at first densely adpressed-sericeous on both surfaces, but in time becoming subglabrous above, though remaining softly silky and greyish below; base of leaf usually cuneate, apex acute, margins commonly entire or at most remotely and obscurely toothed; stipules wanting or soon deciduous. Catkins appearing with the leaves in May and June, the females similar to those of *S. caprea* var. *caprea,* but often distinctly pedunculate, with conspicuous bracts and a densely pubescent peduncle; catkin-scales usually small and much shorter than the ovary-pedicels; ovary narrowly flask-shaped with an elongate, slender neck, densely sericeous-pilose; style indistinct; stigmas oblong, shortly retuse or sometimes cleft into 2 linear lobes. Male catkins closely similar to those of *S. caprea* var. *caprea,* but usually rather smaller and more scattered.

At first sight so distinct by reason of its silky, entire, exstipulate leaves that collectors have more than once mistaken it for *Salix lanata,* or supposed that it was a hybrid between *S. caprea* and *S. lanata* or *S. lapponum.* On closer inspection most of the characters which make var. *sphacelata* so distinct are found to vary, and the number of transitional variants, linking var. *caprea* and var. *sphacelata,* is such that it is difficult to know what rank to give the variant. Nor is it known if var. *sphacelata* is genetically distinct, or if it would retain its distinctive characters in cultivation, though one suspects that it would.

S. caprea var. *sphacelata* was first described (as a species) by J. E. Smith in 1804 (*26* p. 1066), from specimens collected "at Finlarig at the head of Loch Tay" by the Rev. John Stuart (specs. in Lightfoot's herbarium, Kew). It has since been collected many times in highland areas of Perthshire, Argyll, Inverness and Angus, and may prove widespread in the Highlands at altitudes above 300 m (1000 ft). In the circumstances one might expect to find the variety in mountainous areas of northern England, Wales and Ireland, but I have yet to see specimens. Abroad *S. caprea* var. *sphacelata* is locally common in Scandinavia, Russia, and in the central European Alps. It is represented (as *S. caprea*) in the Linnaean herbarium, together with what is here interpreted as typical material of this species.

The epithet *sphacelata* means "gangrened" and alludes to the "remarkably withered or blasted appearance" of the leaf tips, not, however, a normal or diagnostic feature of the variety.

A: leaves and catkins × 2/3; B: larger leaf × 2/3; C: detail of catkin × 6.

Salix caprea var. **sphacelata** 12a

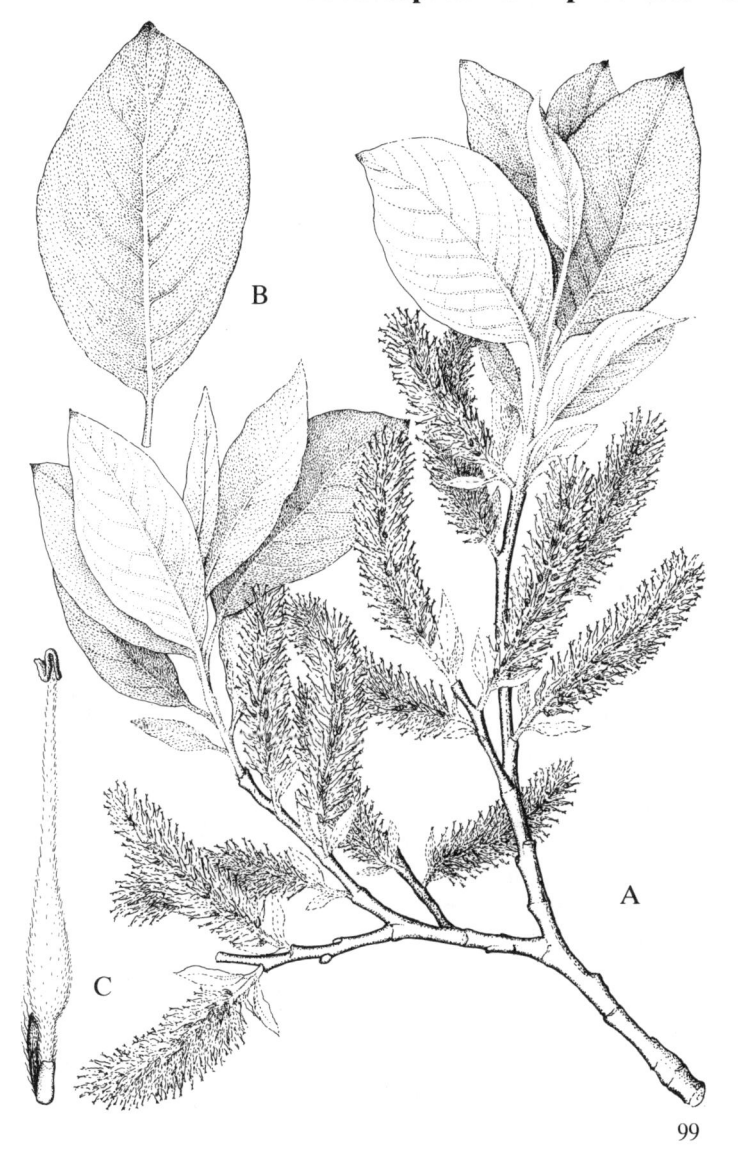

12 × 9. Salix caprea L. × S. viminalis L. = Salix × sericans Tausch ex A. Kerner

Robust bush or small tree to about 9 m high, with lightly fissured bark and spreading branches; twigs yellowish or reddish, at first densely ashy-white, but soon becoming glabrous or subglabrous and rather lustrous; wood of peeled twigs without *striae*. Leaves ovate-lanceolate or lanceolate, usually rather broadly so, about 6–12 cm long, 1.3–3 cm wide, dull green and glabrescent above, densely grey-tomentellous below and very soft to the touch; nervation prominent, often tinged reddish, apex gradually acuminate, base rounded or broadly cuneate, margins usually somewhat undulate, narrowly recurved, with small distant teeth or subentire; petiole 1–1.5 cm long, generally rather stout; stipules narrowly lunate or ear-shaped, acuminate, remotely toothed, sometimes well-developed, but more often small and caducous. Catkins appearing in early spring (March-April), well in advance of the leaves, usually numerous and rather crowded towards the tips of the twigs. Male catkins shortly cylindrical 2–3 cm long, often slightly more than 1 cm wide, sessile or subsessile, with a few, inconspicuous, sericeous bracts; catkin-scales ovate-elliptic, about 2 mm long, 1 mm wide, brownish with a darker, fuscous apex, densely grey-hairy; filaments up to 1 cm long, slender, glabrous; anthers golden-yellow; nectary small, oblong, truncate. Female catkins generally longer and proportionally narrower, up to 5 cm long, usually less than 1 cm wide, ovary narrowly flask-shaped, about 3 mm long, densely grey-sericeous, shortly pedicellate; style generally rather conspicuous, stigmas narrowly oblong, usually longer than style, 2-cleft when mature, with linear lobes.

A common hybrid throughout our area, often apparently spontaneous, but not infrequently an obvious introduction, with the same clone recurring over wide areas, and almost certainly a relic of cultivation, since the hybrid was formerly considered of some value for coarse basket work.

It is possible that the earliest, and correct, name for the hybrid is *Salix × holosericea* Willd. (*Sp. Plant.*, 4 (2): 708 (1806)), despite the fact that many continental authors identify this as *S. cinerea × S.*

A,B: leaves × 2/3; C: stripped twig × 2/3; D: stipule × 4; E: ♂ catkins × 2/3; F: detail × 8; G: ♀ catkin × 2/3; H: detail × 8.

Salix × sericans 12 × 9

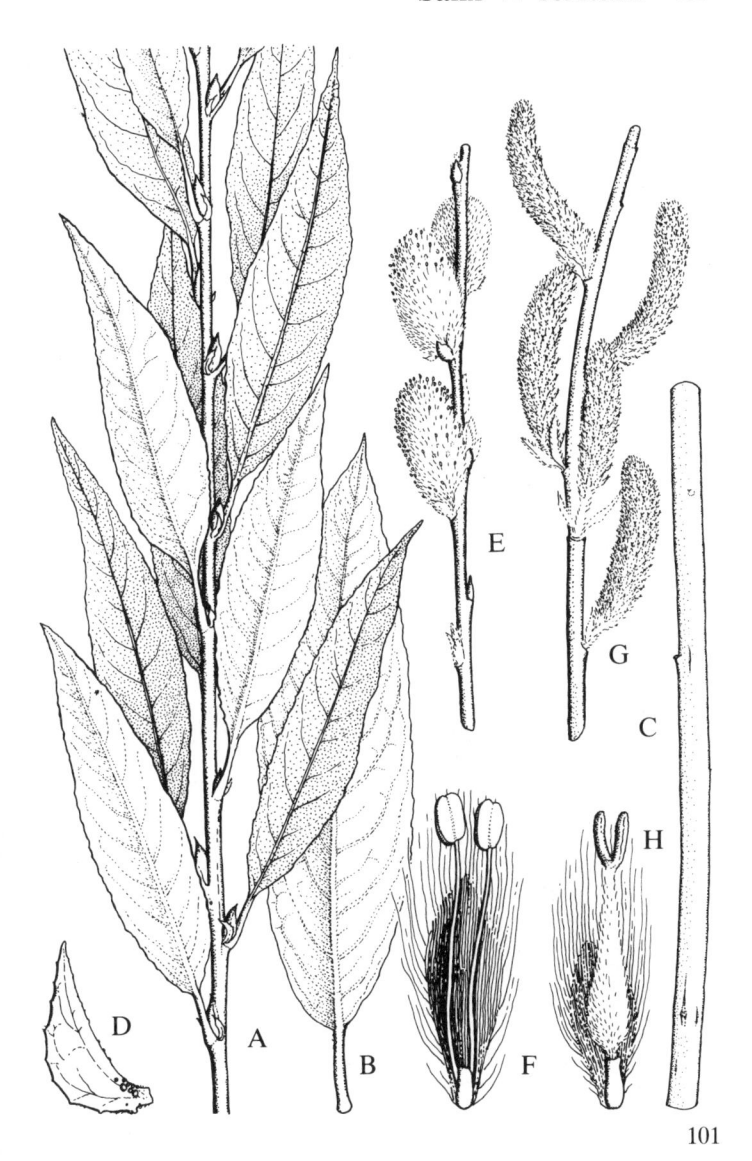

viminalis. Most of the older authors call it *S. mollissima* Sm., but this name is objectionable for two reasons: first, because Smith attributed it to Ehrhart, and it is now agreed that Ehrhart's *S. mollissima* is a *S. triandra-viminalis* hybrid, and secondly because the *S. mollissima* figured by Sowerby in *English Botany* t. 1509 (1805), which was subsequently (1809) to become the basis of *S. smithiana* Willd., appears to be a hybrid between *S. cinerea* and *S. viminalis*.

There is always some difficulty in distinguishing *S.* × *sericans* (*S. caprea* × *S. viminalis*) from *S.* × *smithiana* (*S. cinerea* × *S. viminalis*), nor can it be denied that, however careful the examination, one is generally left with a residue of specimens which refuse to conform to either hybrid, and which are probably *S. caprea* × *S. cinerea* × *S. viminalis*. Allowing for such problems, it is normally possible to identify *S.* × *sericans*, especially in the field, and the principal differences between the two hybrids are set out under *S. cinerea* × *S. viminalis*. It is essential that determinations should be based on mature foliage, since immature leaves of *S.* × *sericans* are virtually indistinguishable from those of *S.* × *smithiana*.

12 × 13. Salix caprea L. × S. cinerea L. = S. × reichardtii A. Kerner

Buchanan-White (*31*) and Linton (*16*) both regarded this hybrid as uncommon, and C. E. Moss (*18*) seems to have been the first to realize that it was much more frequent and widespread than had been supposed, remarking "to our own experience, wherever the putative parents grow together, individuals occur which are with difficulty referred to either the one species or the other. As we find no such difficulty where only one of the species occurs, it is reasonable to suppose that the doubtful plants are of hybrid origin". This is exactly my own experience, and, as mentioned under *Salix caprea*, I conclude not only that the hybrid is very common, but also, because of its polymorphic character, often very puzzling, apparently linking the two parents by an unbroken series of intermediates. Generally the slender, puberulous, dark reddish-brown twigs will indicate the presence of *S. cinerea*, while the

---

A,B: leaves × 2/3; C: stipule × 3; D: ♂ catkins × 2/3; E: detail × 8; F: ♀ catkins × 2/3; G: detail × 8.

Salix × reichardtii   12 × 13

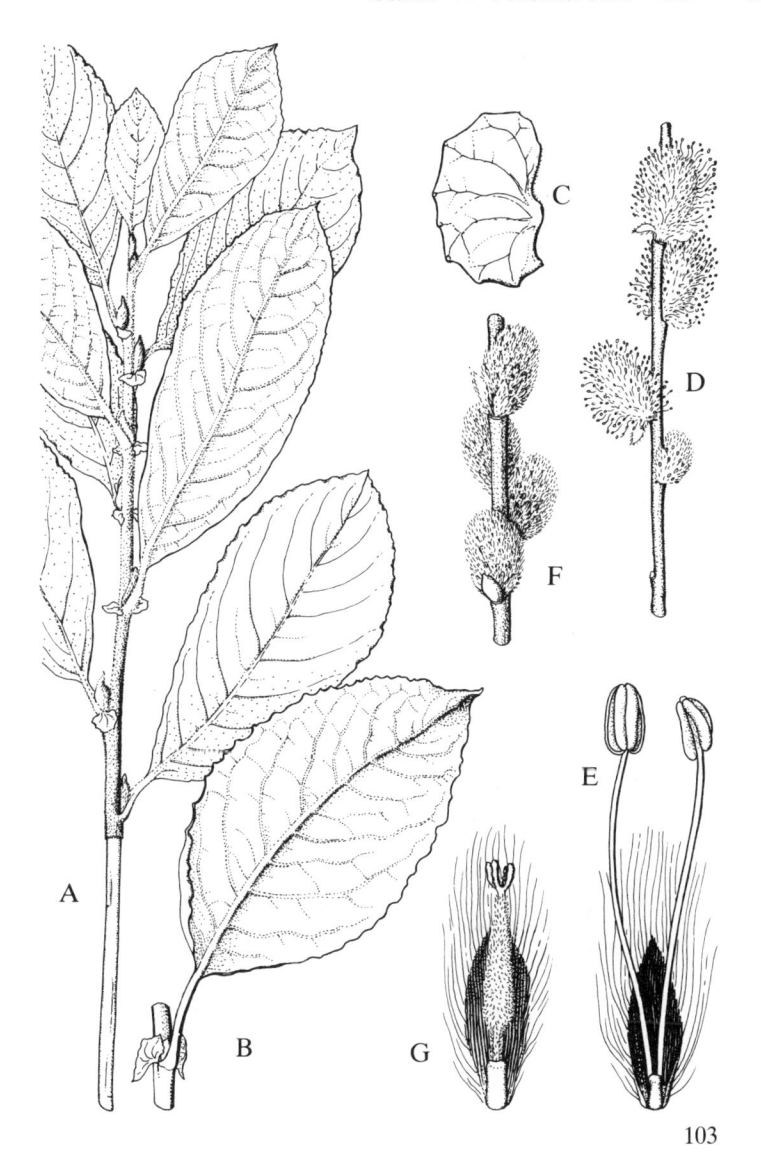

broad, rather rugose-margined leaves, and the soft indumentum of the leaf-undersurface, reveal *S. caprea*. The catkins of the hybrid are generally smaller and more narrow than those of *S. caprea*, though this is a matter of degree, and the wood of the twigs is often marked with a few of the longitudinal *striae* characteristic of *S. cinerea*. Pure *Salix caprea* is, I am certain, distinctly less common than is generally supposed; it is a bush of woodland margins, often on well-drained ground; where the woods have been felled, or the habitat otherwise disturbed, the hybrid and *S. cinerea* tend to replace *S. caprea,* which may, on occasion, survive only as an isolated bush, obviously older than the surrounding scrub, and clearly a relic of a past association.

Most of the hybrid material in our herbaria must consist of *S. caprea* × *S. cinerea* ssp. *oleifolia,* but I doubt if this could easily be distinguished from *S. caprea* × *S. cinerea* ssp. *cinerea* on morphological characters, though it is true that the presence of *S. cinerea* ssp. *oleifolia* is sometimes evident, especially late in the season, from the occurrence of a few scattered rufous hairs on the undersurface of the leaf. At the same season the leaf-undersurface of the hybrid generally becomes sparsely pubescent and often rather harsh to the touch, while that of *S. caprea* remains softly hairy.

13. Salix cinerea L.

A tall shrub or small tree, usually less than 10 m high, generally much branched from the base, but sometimes with a distinct trunk; bark dark grey-brown, becoming fissured with age, branches spreading to form a broad, rounded or flattened crown; twigs dark reddish-brown, at first densely pubescent, and sometimes remaining so at least until the end of the first year, or becoming glabrous or subglabrous; wood of peeled twigs with long, scattered *striae*; leaves very variable, usually obovate or broadly oblanceolate, sometimes oblong or elliptic or suborbicular, 2–9(–16) cm long, 1–3(–5) cm wide, dull grey-green and pubescent above or dark green and sublustrous, ashy grey below, and either softly pubescent or thinly clothed with short, scattered rusty-red hairs; nervation prominent or obscure; apex acute or obtuse, rarely twisted obliquely; base cuneate; margins narrowly recurved, distinctly but irregularly undulate-serrate or less often subentire; petiole usually less than 1 cm long; stipules rather broadly ear-shaped, sometimes persistent and almost as long as the petiole, or small and caducous, or wanting. Catkins appearing in advance of the leaves in March and April, erect, sessile or subsessile at first, shortly pedunculate later, cylindrical or narrowly ovoid, 2–3(–5) cm long, 0.6–1 cm wide; bracts short, inconspicuous, densely hairy; catkin-scales oblong, acute or obtuse, fuscous, densely villose, 2–2.5 mm long, 1.5–2 mm wide. Male catkins with 2 free stamens, filaments glabrous, 5–7 mm long, anthers oblong, yellow, about 1.5 mm long, 0.5–0.8 mm wide; nectary narrowly oblong, about 1–1.5 mm long, truncate. Female catkins often smaller and narrower than the male; ovary 2.5–5 mm long, about 1–2 mm wide, flask-shaped, grey-tomentose; pedicel 2–2.5 mm long, generally much longer than the nectary; style short, or occasionally distinct; stigmas narrowly oblong, about 0.5 mm long, erect or sometimes spreading, subentire or more or less deeply 2-cleft. Capsule up to 10 mm long, 2.5 mm wide.

Our populations of this common Willow can be sub-divided into two tolerably distinct races or subspecies:

3a. Salix cinerea L. ssp. cinerea

Grey Sallow

Generally a much-branched shrub 4–6(–10) m high; twigs at first densely pubescent, tardily glabrescent and often retaining some of their indumentum until the end of their second year; wood conspicuously striate; leaves often broadly obovate or oblong with distinctly undulate-serrate margins; lamina dull green and shortly pubescent above, densely and softly pubescent below at first, but becoming glabrescent or thinly pubescent and ashy-grey with age; stipules often large and persistent.

John Fraser, acting on the advice of the eminent Swedish salicologist, Dr Bjorn Floderus, too hastily concluded that typical *Salix cinerea* could "be dismissed from our minds as not British" (*12* p. 369), with the consequence that, for a decade or more, all our material, previously named *S. cinerea,* was indiscriminately re-identified as *S. atrocinerea* Brot. (*S. cinerea* L. ssp. *oleifolia* Macreight). Now it is acknowledged that typical *S. cinerea* has a wide distribution in the fenlands of southern and eastern Britain, and that, as in Norfolk, it may locally replace the more generally distributed ssp. *oleifolia* (see Petch & Swann, *Flora of Norfolk,* 177 (1968)). The exact limits of the distribution of *S. cinerea* are not at present known, but specimens, which seem to be correctly named, from localities as far apart as Perthshire, Isle of Man and Co. Down (Northern Ireland) suggest that it may be commoner than is generally supposed. In most of its recorded localities, it has been collected in base-rich fens and marshes at low altitudes, and it would appear to be more restricted ecologically than ssp. *oleifolia.* While some populations, notably in Norfolk, Cambridge, Lincs. and Hunts., are uniformly distinct and identifiable by the characters given above, it must be admitted that as one travels further west, the differences between the two subspecies become less obvious, though whether this is clinal, or through hybridization, is not certainly known. It is also true that Continental specimens of *S. cinerea* ssp. *cinerea* are, on the whole, larger-leaved and more densely ashy-pubescent than all but the most extreme of our specimens. On the Continent, ssp. *cinerea* has a very wide distribution, extending from Scandinavia through Holland, eastern France and Italy eastwards to Siberia.

Salix cinerea ssp. *cinerea* can be confused with the relatively common *S. aurita* L. × *S. cinerea* L. ssp. *oleifolia* Macreight, but this has more slender, darker-coloured, less ashy twigs, leaves usually with obliquely twisted tips, and smaller catkins.

13b. Salix cinerea L. ssp. oleifolia Macreight

Rusty Sallow

Often a taller shrub than *S. cinerea* L. ssp. *cinerea* and not infrequently a small tree 10–15 m high with a well-developed trunk and irregularly fissured bark ("Truly arboreous, and if allowed to grow, as tall as a common Crab-tree", Smith, *English Flora,* 4: 219 (1825)); twigs soon glabrescent and dark reddish-brown; wood usually rather weakly striate; leaves often oblanceolate or narrowly oblong, not infrequently entire or remotely and obscurely undulate-

---

A,B: leaves × 2/3; C: stipule × 3; D: ♂ catkins × 2/3; E: detail × 8; F: ♀ catkins × 2/3; G: detail × 8.

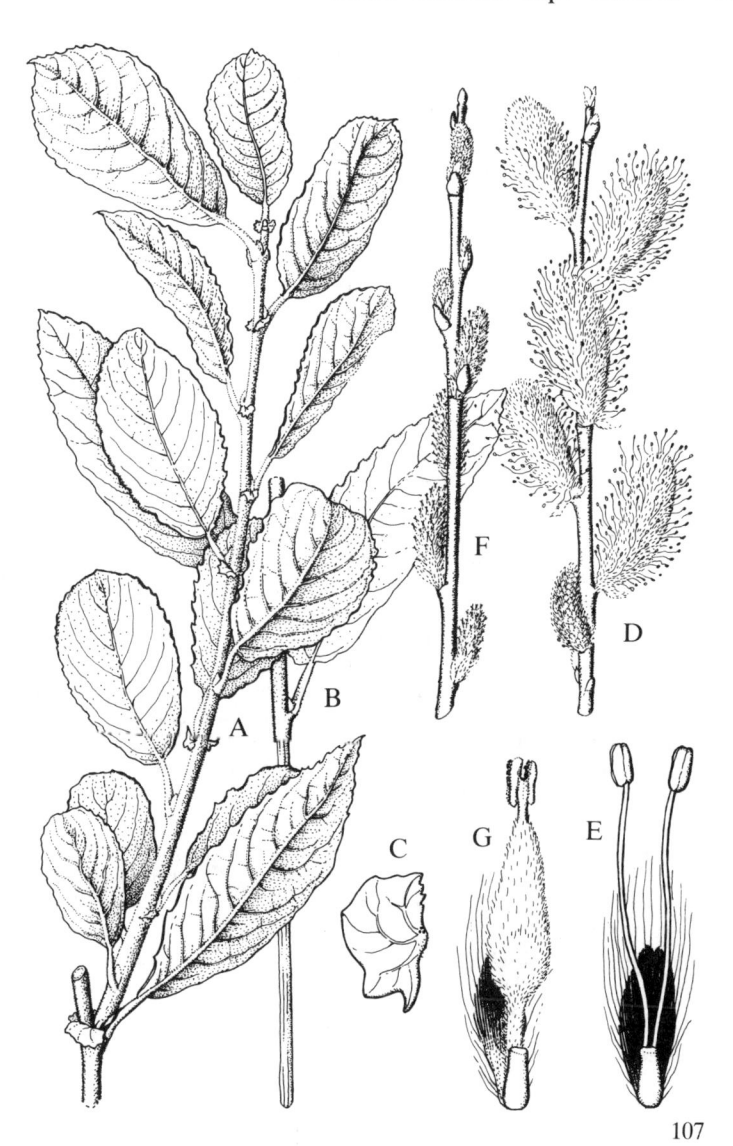

Salix cinerea ssp. cinerea 13a

serrate, but not always distinguishable in shape from those of ssp. *cinerea;* lamina rather lustrous dark green above, ashy-grey below but usually with a sparse indumentum of short, stiff rusty hairs and minute blackish glands; the rustiness is particularly conspicuous towards the tips of the shoots in late summer and autumn; stipules commonly small and caducous. Catkins as in ssp. *cinerea.*

Credit for recognizing this Willow in Britain must go, in the first place, to James Dickson (1738–1822) who communicated specimens—of undisclosed origin—to his friend the assiduous salicologist Sir J. E. Smith. The willow was described by Smith as *Salix oleifolia (26* p.1065), an unfortunate choice of epithet since it was antedated by *S. oleifolia* of Villars (1789), quite a different plant. Regarded as a species, Smith's *S. oleifolia* must in consequence be re-named *S. atrocinerea* Brot. (1804), but at subspecific level (the status now generally agreed as appropriate) *S. atrocinerea* Brot. becomes *S. cinerea* L. ssp. *oleifolia* Macreight (*Manual of British Botany,* 212 (1837)) and not *S. cinerea* L. ssp. *atrocinerea* (Brot.) Silva et Sobrinho (1950). As a variety, it is *S. cinerea* L. var. *oleifolia* Gaudin (*Flora Helvetica,* 6: 242 (1830)). The fact that the names appear at so many ranks indicates the element of uncertainty which has always clung to the taxon. Smith's decision to regard it as a distinct species was never wholeheartedly approved by his British colleagues, but, even so, the name *S. oleifolia* continued to appear in British Floras until 1843, when Babington reduced it to varietal status; by 1850 (Hooker & Arnott, *British Flora,* 392) it had become a straight synonym of *S. cinerea,* and was very largely forgotten for the next forty years, until Buchanan White (31) once again drew attention to the "absence of exact identity" between British and Continental examples of *S. cinerea,* and correctly concluded that the rusty-hued British Willow was *S. atrocinerea* Brot., though he was unwilling, without further study, "to ascribe to the British form varietal or subspecific rank, though it may be that further investigations will show that it is worthy of such". E. F. Linton (*16*) regarded *S. atrocinerea* Brot. as a synonym of *S. cinerea* L., but revived *S. oleifolia* Sm. as a variety of *S. cinerea*; he did not comment on Buchanan White's analysis of British and Continental material of the species, and evidently attached little importance to the distinctions. The whole question was re-opened by Fraser in

---

A: leaves × 2/3; B: stripped twig × 2/3; C: stipule × 3; D: ♂ catkins × 2/3; E: detail × 8; F: ♀ catkins × 2/3; G: detail × 8.

Salix cinerea ssp. oleifolia 13b

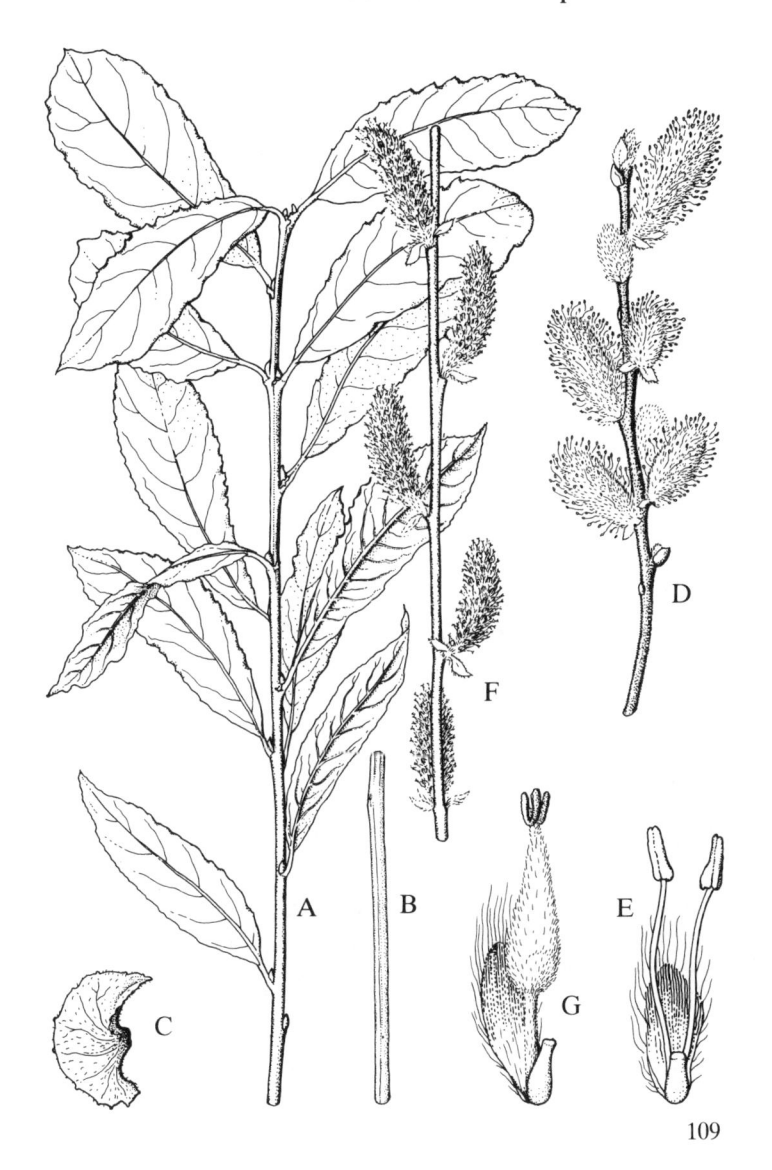

1933 (*12*) when, following Floderus, *S. cinerea* was dismissed as non-British, and *S. atrocinerea* became, for a short period, the sole British representative of the aggregate. As noted under *S. cinerea* ssp. *cinerea*, this extreme view was soon challenged, and it is now generally agreed that both *cinerea* and *atrocinerea* are to be found in this country.

At best the distinctions between the two are somewhat elusive, and certainly not of such moment as to justify separation at species level, but that there is some difference between the pubescent, ashy-leaved *cinerea* and the harsh, lustrous, rusty-leaved *oleifolia* cannot be denied; the fact that the two have distinct geographical distributions and ecological preferences reinforces the conclusion that they are most satisfactorily ranked as subspecies.

S. cinerea ssp. *oleifolia* is quite the commonest of our Willows, and is to be found in local abundance everywhere in Britain and Ireland except in Norfolk and adjacent counties of East Anglia, where it tends to be replaced by ssp. *cinerea*. Smith's comment that "it [ssp. *oleifolia*] proves to be not uncommon in hedges and coppices in various parts of Norfolk" (*English Botany*, 20: t. 1402 (1805)) has not been confirmed by recent investigations. On the Continent ssp. *oleifolia* is seemingly restricted to W. France, Spain and Portugal. It is found in a much wider range of habitats than ssp. *cinerea*, growing on acid or basic soils by streamsides, edges of bogs and marshes, moist woodland margins and hedgerows, from sea-level to 600 m (2000 ft) in Perthshire (Wilson, *North-west Naturalist*, 26, Suppl. (1956)). On very acid soils it is commonly replaced by *S. aurita* L.

13 × 16. Salix cinerea L. × S. phylicifolia L. − Salix × laurina Sm.

Erect, lax shrub 2–6 m high; young shoots thinly pilose with spreading, straight and crisped hairs, soon becoming glabrous; older twigs rich lustrous brown; leaves elliptic, oblong or narrowly obovate, 4–10 cm long, 1.8–3.5 cm wide, dark rather lustrous green and glabrous or subglabrous above, conspicuously glaucous and thinly pubescent along the midrib and nerves below, apex shortly

---

A: leaves × 2/3; B: ♀ catkins × 2/3; C: detail × 8.

Salix × laurina 13 × 16

acute, occasionally obliquely twisted, margins very narrowly recurved, entire, subentire or very shortly and rather bluntly serrate; petiole normally short and stout, seldom more than 1 cm long; stipules small, spreading, bluntly ear-shaped, shortly denticulate, soon deciduous. Catkins narrowly cylindrical, 3–5 cm long, 0.8–1 cm wide, appearing a little in advance of the leaves in April and May, rather evenly spaced along the year-old twigs; peduncles densely villose, short, but usually distinct, up to 1 cm long; bracts lanceolate, spreading, 6–10 mm long, 2–4 mm wide, subglabrous above, densely sericeous-pilose below; catkin-scales ovate-oblong, about 2–3 mm long, 1–1.5 mm wide, brownish, usually with a fuscous apex, obtuse or subacute; female catkins with numerous, crowded, erect, narrowly flask-shaped, conspicuously white-tomentose ovaries, at first 3–4 mm long, 1–1.5 mm wide, distinctly accrescent in fruit; pedicel short, pubescent, at first about half as long as catkin-scale, lengthening with age; style generally distinct, about 1 mm long; stigmas linear-oblong, usually cleft to base, sometimes undivided. Male catkins not seen, but one specimen (E. F. Linton 52 from Thornhill, Dumfriesshire) with androgynous catkins, has 2 glabrous filaments about 6 mm long, and small, yellow anthers.

Smith (*Trans. Linn. Soc.,* 6: 122) does not record the orgin of his *Salix laurina* beyond noting that it came to him from James Dickson, who most probably collected it in Scotland, where it is not uncommon. E. F. Linton is correct in regarding *S. wardiana* Leefe ex F. B. White and *S. tenuior* Borrer (not *S. tenuifolia* as given by Linton) as synonyms. The last-named has narrower, more obovate leaves than is usual in *S.* × *laurina,* and comes nearer to *S. cinerea* (ssp. *oleifolia*) than to *S. phylicifolia.*

Smith renamed the plant as *S. bicolor,* but it is not *S. bicolor* Ehrh., nor is the plate in *English Botany,* t. 1806, very good, since the leaves have been drawn from abnormal sucker or coppice shoots, and look more like *S.* × *calodendron* than *S.* × *laurina,* though specimens at Kew sent by Smith to Dawson Turner in 1803 prove that the plant illustrated as *S. bicolor* in *English Botany,* is unquestionably *S.* × *laurina.* Why Smith should regard *S.* × *laurina* as frequent in Norfolk is hard to understand, since it is a northern plant, locally plentiful from Yorkshire northwards, and planted in Warwickshire and in the Shetlands. The uniformity of many populations of *S.* × *laurina,* and the general absence of male plants, suggest that in many areas it is represented by planted specimens derived from one original stock. Certainly the *S.* × *laurina* (or more usually *S.* × *wardiana*) of botanic gardens is very homogeneous, and is clearly propagated clonally.

S. × *laurina* is easily recognised by its firm leaves, shining green above, strongly glaucous below, also by its densely white-tomentose ovaries, with well-marked styles. It is certainly *S. cinerea* × *S. phylicifolia* as maintained by E. F. Linton, and the *S. laurina* of Heribert Nilsson (*Lunds Univ. Årsskr.* N.F., avd. 2, 24 No. 6 (1928)) though no doubt *S. caprea* × *S. viminalis* as the author states, is not the *S. laurina* of Smith.

Some forms of *S. cinerea* L. ssp. *oleifolia* Macreight can be misleadingly similar to *S.* × *laurina*, but generally lack the contrasting leaf-surfaces, white-tomentose ovaries, and well-marked styles of the hybrid.

S. × *laurina* occurs in Scandinavia (where it has also found its way into Botanic Gardens). Its spontaneous occurrence elsewhere in Europe is more doubtful.

3 × 6. Salix cinerea L. × S. purpurea L. = Salix × sordida A. Kerner

An erect or spreading, much branched shrub, up to 5 m high; twigs rather slender, at first thinly pubescent, but soon glabrous and lustrous reddish-brown. Leaves oblong-elliptic or obovate, 4–7 cm long, 1.5–2.5 cm wide, at first covered with a pallid tomentum, but very soon glabrous and shining green above, ashy-grey below and thinly pubescent, or glabrous with a few hairs along midrib and nerves, apex acute or cuspidate, margins irregularly and minutely serrate with glandular teeth, or subentire; nervation often rather conspicuous and reticulate; petiole short, seldom more than 5 mm long; stipules narrowly ear-shaped, acute, with glandular-denticulate margins, caducous or often wanting. Catkins appearing before the leaves in late March or April, erect or erecto-patent, cylindrical, sessile or subsessile, about 1.5–3 cm long, 0.5·0.7 cm wide; bracts lanceolate, 3–8 mm long, 1.5–3 mm wide, glabrous above, densely sericeous-hairy below; catkin-scales ovate or obovate, acute or obtuse, fuscous or reddish towards base, rather densely hairy, about 1.5 mm long, 1 mm wide. Male catkins with 2 free or partly connate stamens; filaments glabrous; anthers oblong, about 0.7 mm long, at first orange-red, but turning yellow on dehiscence; nectary narrowly oblong, truncate. Female catkins with densely grey-tomentose, very shortly pedicellate, crowded, narrowly ovoid ovaries 1.5–2 mm long, 0.8 mm wide, the pedicel shorter than the oblong nectary; style absent or almost absent; stigmas shortly oblong, erect or somewhat spreading, usually entire.

Although C. E. Moss (*18* p. 67) says: "where *S. cinerea* and *S. purpurea* grow together, intermediates between them appear to be not uncommon" few of these intermediates are to be found in herbaria, and *Salix* × *sordida* must be reckoned a rare hybrid. The earliest record, under the illegitimate name *S. pontederana* Willd., is sometimes attributed to the Rev. J. E. Leefe, who collected what was supposed to be *S. cinerea* × *S. purpurea* by the R. Coquet near Rothbury, Northumberland, some time before 1872, but, as already noted by E. F. Linton, Leefe's specimens (*Salictum Exsicc.* Fasc. III, 20. 59) belong to the rare *S. aurita* L. × *S. purpurea* (*S.* × *dichroa* Doell). Other records, possibly genuine, are from E. Gloucestershire, Northampton and Westmorland, and I have seen satisfactory herbarium specimens from Dumfries and Perthshire. Regarding the last-mentioned, Buchanan White (*31* p. 450–1), writes: "In the Woody Island near Perth, where both *S. purpurea* and *S. cinerea* abound, *S. sordida* appears to be not uncommon. But while forms quite intermediate between the parent species occur, the majority of individuals are in character much nearer *S. cinerea*—so near, in fact, that in the absence of flowers, some of them would with difficulty be discriminated from *cinerea* otherwise than as slight modifications". Buchanan White also remarks that, although the filaments are usually to some degree connate in *S.* × *sordida*, "frequently the filaments are united for a little way above the base only; but in the same catkin free, slightly united and more distinctly connate stamens may all be found".

In addition to the Woody Island population, White notes *S.* × *sordida* from "Dalmarnock, on the Tay above Dunkeld, and on the Tay below Perth".

The name *S. pontederana* appears spasmodically in literature before 1870, but such earlier references are either spurious, or based upon cultivated material distributed by Forbes from the *Salictum Woburnense*.

S. × *sordida* has a wide distribution in Europe.

A: leaves × 2/3; B: stipule × 12; C: ♂ catkins × 2/3; D: detail × 12; E: ♀ catkins × 2/3; F: detail × 12.

Salix × sordida   13 × 6

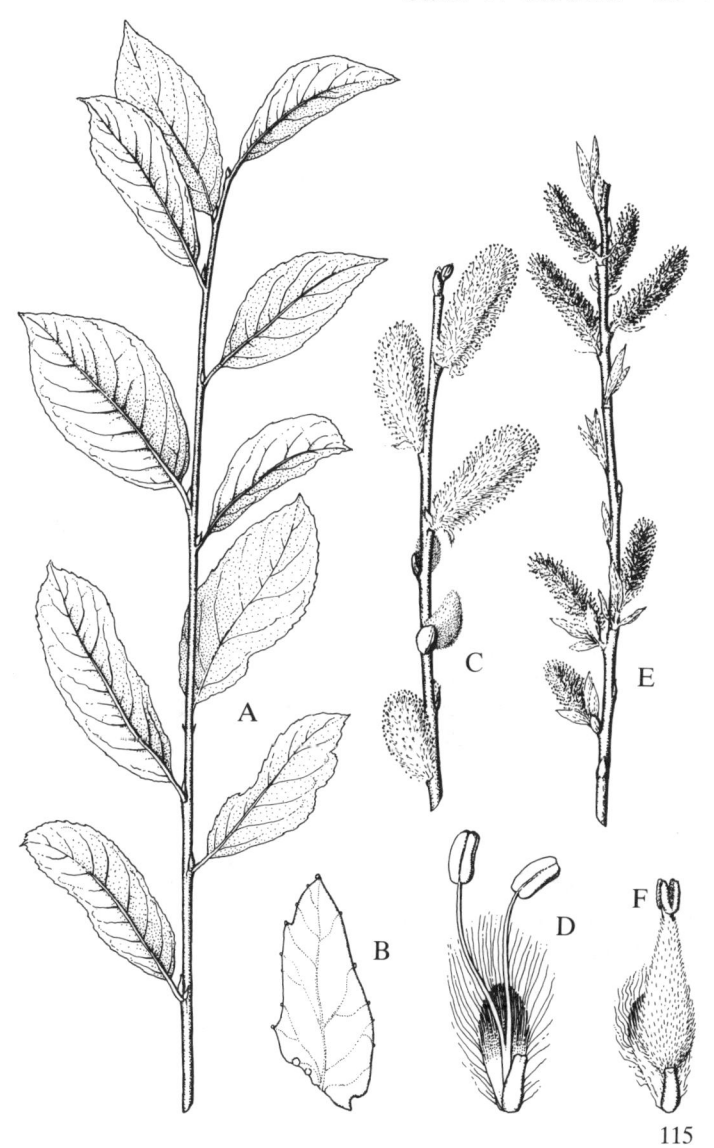

13 ×17. Salix cinerea L. × S. repens L. = Salix × subsericea Doell

A low, sprawling shrub, generally less than 1 m high, with slender, dull, dark purple-brown glabrescent twigs; the leaves are narrowly oblong or ovate-elliptic in outline, 2–4.5 cm long, 0.8–2.3 cm wide, bright green and very thinly adpressed-hirsute or subglabrous above, ashy-grey below, and sometimes conspicuously sericeous-pilose but sometimes almost glabrous, generally with a straight or oblique, acute or subacute apex, and narrowly recurved margins. The petioles are rarely more than 5 mm long, and the stipules, generally fairly obvious and persistent in the closely similar *S.* × *ambigua (S. aurita* × *S. repens)*, are invariably small and soon deciduous in *S.* × *subsericea*. The catkins appear a little in advance of the leaves, generally in late April or early May, and are sessile or very shortly pedunculate, with a few, small, basal, sericeous bracts; the males have not yet been collected in Britain; the females are about 2–3 cm long, 0.6–0.8 cm wide, with dark, fuscous-tipped, oblong, obtuse or subacute, hairy scales, and very shortly stalked or subsessile, narrowly flask-shaped or subcylindrical, thinly grey-pubescent ovaries 2–3.5 mm long; the styles are short, but often distinct, and the stigmas are generally 2-cleft, but sometimes subentire.

Although recorded from more than 10 vice-counties, and with a wide distribution on the Continent, *S.* × *subsericea* is, in my experience, a very uncommon hybrid, and many of the records are questionable, being either misidentifications of the relatively common *S. aurita* × *S. repens,* or, more often than not, forms (or states) of *S. cinerea,* dwarfed and flattened by exposure. I have, however, seen genuine material of the hybrid from Merioneth and from dunes at Tents Muir in Fife. These, and presumably, the deliberately manufactured *S. cinerea* × *S. repens* distributed by W. R. Linton, from Shirley, S. Derbyshire, in 1898, are derived from *S. cinerea* L. ssp. *oleifolia* Macreight, rather than from *S. cinerea* ssp. *cinerea*. The Linton specimens are, indeed, so similar to forms of *S. cinerea* ssp. *oleifolia* that their hybridity might, in other circumstances, be called in question. According to Håkansson (*15*) male plants are rare in the F_1 generation.

A: leaves × 2/3; B: stipule × 8; C: ♀ catkins × 2/3; D: detail × 8.

Salix × subsericea 13 × 17

13 × 9. Salix cinerea L. × S. viminalis L. = Salix × smithiana Willd.

A robust shrub or small tree to about 9 m high with lightly fissured bark and spreading branches; twigs dark reddish-brown, densely pubescent at first and generally remaining so throughout the first year; wood of peeled twigs generally with a few striae. Leaves narrowly lanceolate-acuminate, 6–11 cm long, 0.8–2.5 cm wide, dull green and glabrescent above, shortly grey-sericeous below, becoming subglabrous and greenish with age; nervation usually prominent, but not conspicuously so, sometimes reddish; apex acuminate, base cuneate, margins narrowly recurved, generally serrulate with numerous small teeth; petiole 5–13 mm long, rather stout; stipules lunate or ear-shaped, acuminate, often well developed and rather persistent, occasionally with several small basal appendages. Catkins appearing in early spring (March–April) in advance of the leaves, usually numerous and rather crowded towards the tips of the twigs. Males shortly ovoid-cylindrical, 2–3 cm long, 1–1.2 cm wide, sessile or subsessile with a few, inconspicuous, sericeous bracts; catkin-scales ovate-elliptic, acute, 2–2.5 mm long, 1 mm wide, brownish with a darker, fuscous apex, densely grey-hairy; filaments up to 8 mm long, slender, glabrous; anthers golden-yellow; nectary small, oblong, truncate. Female catkins often longer and proportionally narrower than the male, commonly 3 cm or more long but usually less than 1 cm wide; ovary narrowly flask-shaped, about 3 mm long, densely grey-sericeous, shortly but distinctly pedicellate, style generally developed, stigmas narrowly oblong, about as long as the style, often cleft into 2 linear lobes at maturity.

One of the commoner *Salix* hybrids, probably occurring throughout Britain and Ireland, but generally less frequent than *S. caprea* × *S. viminalis*, which it greatly resembles. Although the uniformity of some local populations suggests deliberate introduction, *S.* × *smithiana* is, unlike *S.* × *sericans,* often spontaneous. Male plants are much less common than females, and Håkansson (*15*) comments on the rarity of males in artificial hybrids between *S. cinerea* and *S. viminalis*.

Although it is sometimes difficult to distinguish *S. caprea* × *S.*

A: leaves × 2/3; B: stripped twig × 2/3; C,D: stipules × 4; E: ♂ catkins × 2/3; F: detail × 8; G: ♀ catkins × 2/3; H: detail × 8.

Salix × smithiana 13 × 9

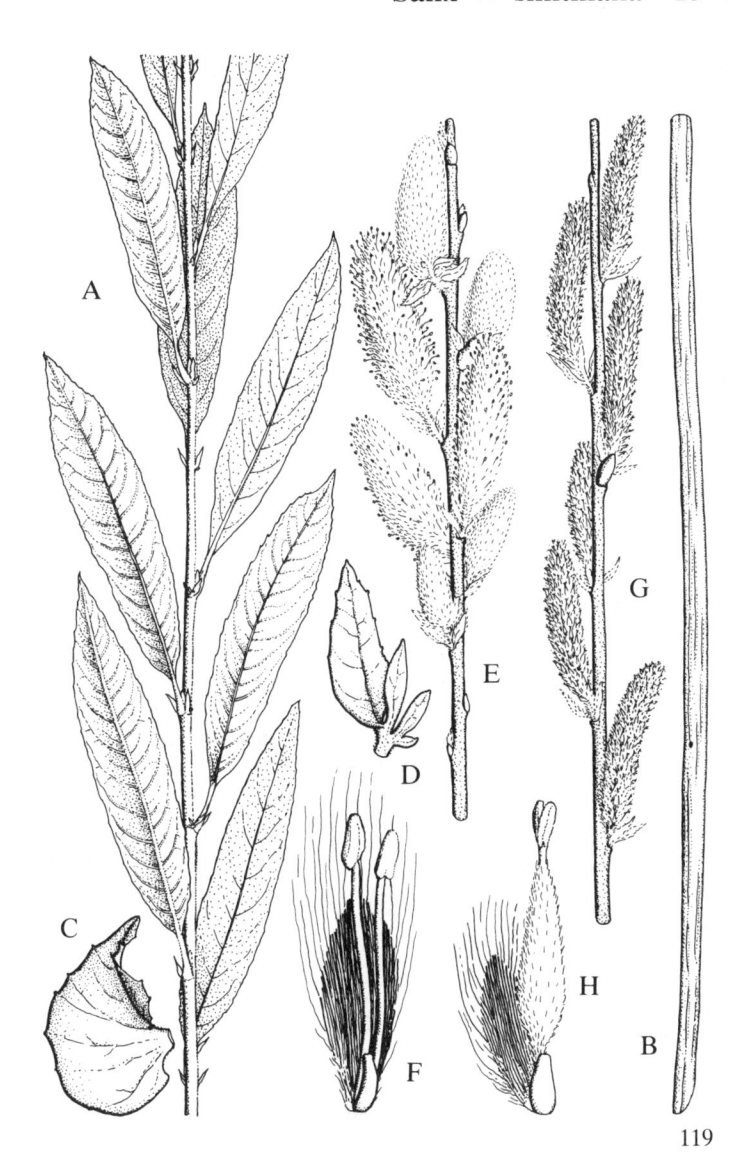

viminalis from *S. cinerea* × *viminalis,* the following characters generally serve to separate the two: *S. caprea* × *S. viminalis* has pale twigs, usually yellowish or tinged red, with the underlying wood smooth (not striate) when the bark is peeled off. The leaves are softly and sometimes densely tomentellous below, with very prominent reticulate nervation. *S. cinerea* × *S. viminalis* has reddish brown twigs, with the underlying wood usually striate when the bark is peeled off. The leaves are shortly tomentellous, sometimes almost glabrous below at maturity and the nervation is much less prominently reticulate. Leaf size is also a useful pointer, the leaves of *S. caprea* × *S. viminalis* being usually larger, broader and with a more rounded base than in *S. cinerea* × *S. viminalis,* but there is considerable overlap in these characters, so that they are not always reliable.

Salix × *smithiana* is widely distributed on the Continent, where it is presumably most often *S. cinerea* ssp. *cinerea* × *S. viminalis*; in our area some hybrids, especially those from East Anglia, will have this parentage, others, from further west, will mostly be *S. cinerea* ssp. *oleifolia (S. atrocinerea)* × *S. viminalis,* but it is questionable if the two infraspecific hybrids can always be distinguished. Typical *S.* × *smithiana* appears to be *S. cinerea* ssp. *cinerea* × *viminalis,* while the hybrid with *S. cinerea* ssp. *oleifolia* is *S.* × *smithiana* Willd. nothovar. *ferruginea* (G. Anderson ex Forbes) Leefe (*S. ferruginea* G. Anderson ex Forbes, *Salict. Woburn.,* t. 128 (1829)). The same hybrid has also been named *S.* × *chouardii* Chassagne et Görz (1931). Where the characteristic rufous indumentum of *S. cinerea* ssp. *oleifolia* reappears in the hybrid one can be sure of the parentage, but, in my experience, the rufousness is often absent, even in circumstances where one is reasonably sure that *S. cinerea* ssp. *oleifolia* is one of the parents.

14. Salix aurita L.

Eared Willow

A much-branched shrub, generally less than 2.5 m high with numerous intercrossing branches, twigs slender, dark reddish-

---

A: leaves × 2/3; B: stripped twig × 2/3; C: stipule × 3; D: ♂ catkins × 2/3; E: detail × 12; F: ♀ catkins × 2/3; G: detail × 12.

Salix aurita 14

brown, at first shortly pubescent, soon becoming glabrous or subglabrous; wood of twigs with numerous, prominent *striae*; leaves generally obovate or oblong-obovate, 2–6 cm wide, rugose, dark, dull green above, ashy-grey below and generally softly pubescent, with prominent nervation; apex rounded or shortly acute, commonly with an obliquely twisted apiculus; base cuneate, usually tapering to a short, erecto-patent petiole 3–8 mm long; margins normally undulate-serrate, with short, rather irregular teeth, occasionally subentire, stipules conspicuous, broadly ear-shaped, persistent, with undulate-serrate margins. Catkins appearing in advance of the leaves in April and early May, erect, sessile, shortly cylindrical, 1–2 cm long, 0.7–0.8 cm wide; bracts short, densely silvery-hairy; catkin-scales oblong, acute or obtuse, fuscous or dark reddish, thinly or densely hairy, 1.5–2 mm long, 0.7–0.8 mm wide, those of the female flowers commonly narrower and more acute than the males. Male catkins with 2 free stamens, filaments glabrous or sometimes hairy near the base, 5–7 mm long; anthers oblong, yellow, generally less than 1 mm long and 0.7 mm wide; nectary narrowly oblong, slightly tapering upwards, about 0.5 mm long, truncate at apex. Female catkins often rather longer than the male, ovary narrowly flask-shaped, 1.5–2 mm long, 0.5 mm wide, densely grey-tomentose, pedicel distinct, frequently 2 mm long, and often longer than the catkin-scale; style very short or wanting; stigmas shortly oblong, erect or slightly spreading, generally entire. Capsule up to 8 mm long and 2 mm wide.

A locally common shrub of acid heaths, woods and moorlands, widely distributed in our area, and often replacing, or almost replacing, *Salix cinerea* ssp. *oleifolia* in the more barren parts of Scotland, Ireland and Wales. It is rare over much of the Midlands and East Anglia, and is generally a sure pointer to marginal agricultural land. Though common on the lower slopes of most Scottish and Irish mountains, it has not been recorded at altitudes above 800 m (2600 ft) (Atholl). Outside Britain and Ireland, *Salix aurita* has a wide distribution in central and northern Europe, eastwards to the limits of European Russia. It has been recorded from Turkey and N. Iran, but with questionable accuracy.

On the whole, *S. aurita* is not a very variable species: dwarfed states, growing in very exposed places, have been distinguished as var. *minor* Anderss., and the tall, woodland state, with thin leaves which are green and glabrescent on both surfaces, answers to var. *nemorosa* (Fr.) Anderss. subvar. *virescens* Anderss. But neither variant is of more than trifling taxonomic significance. On the

Continent, much of the material from Scandinavia and northern Europe has larger, less rugose, more acute leaves (more like those of *S. cinerea*) and it may be that the aggregate species includes more than one geographical race.

The species is particularly subject to attack by the gall-midge *Rhabdophaga rosaria,* and the quaint "Rose galls", or terminal rosettes of deformed leaves, are conspicuous in winter, persisting on the bushes long after the leaves have been shed.

14 ×13. Salix aurita L. × S. cinerea L. = Salix × multinervis Doell

An erect, much-branched shrub or small tree, generally less than 5 m high; twigs rather slender, at first closely grey-pubescent, becoming glabrous or subglabrous and dark reddish-brown with age, underlying wood marked with numerous, conspicuous striae. Leaves oblong or obovate, 1.5–2 cm long, 0.7–2.5 (–3.5) cm wide, dark dull green above, ashy-grey or rusty below, at first densely and softly pubescent, becoming subglabrous with age, but generally with the undersurface remaining pubescent, and frequently soft to the touch, even at maturity, apex obtuse or shortly acute, sometimes obliquely twisted, margins often strongly undulate-serrate, nervation generally prominent below; petiole short, usually about 3–6 mm long; stipules commonly conspicuous and persistent, ear-shaped, pubescent, somewhat rugose. Catkins appearing before the leaves in April and early May, erect or spreading, sessile, shortly cylindrical, 1.5–2.5 cm long, 0.7 0.8 cm wide; bracts short, densely silvery-hairy; catkin-scales oblong, obtuse or subacute, fuscous or reddish, usually rather densely hairy, 1.5–2 mm long, 0.7–1 mm wide. Male catkins with 2 free stamens, filaments glabrous or often thinly hairy towards the base; anthers yellow or sometimes tinged red; nectary oblong, truncate. Female catkins with densely grey-tomentose, distinctly pedicellate, flask-shaped ovaries, the pedicel usually rather longer than the oblong nectary; style short, sometimes almost absent; stigmas oblong, erect and spreading, occasionally 2-cleft.

A very common hybrid, found wherever the parent species occur, and not infrequently forming an unbroken series of intermediates between one parent and the other. Generally, however, it is distinguishable from *Salix cinerea,* or, in our area, from *S. cinerea* ssp. *oleifolia,* by the prominent, persistent stipules and rather wrinkled, dull, dark green leaves which tend to remain softly, but thinly, pubescent below. *S. aurita* is a smaller, more twiggy bush, with more slender twigs, and smaller, more wrinkled leaves.

On the Continent, where *S.* × *multinervis* is also common, the great majority of hybrids are between *S. aurita* and *S. cinerea* ssp. *cinerea,* whereas, in our area, this, the typical hybrid, is recorded only from Norfolk (Petch & Swann, *Fl. Norfolk,* 177 (1968)) though the reference to "the frequent hybrid, *S. aurita* × *cinerea,* with its characteristic ferrugineous indumentum on the lower leaf-surface" is puzzling, since this, I should have thought, was characteristic only of *S. cinerea* ssp. *oleifolia* (*S. atrocinerea* Brot.) and its hybrids. If this latter is distinguished as a species from *S. cinerea,* then the hybrid *S. atrocinerea* × *S. aurita* is *S.* × *charrieri* Chassagne (in Guétrot, *Pl. Hybr. France,* III & IV; 113 (1929)). In the absence, however, of supplementary external evidence, I think it might be very difficult to identify the *cinerea* parent with certainty. Under current rules of nomenclature, those who regard *S. atrocinerea* as a subspecies of *S. cinerea* should, probably, rank *S.* × *charrieri* as a nothosubspecies of *S.* × *multinervis*. The 1978 *Code* (Rec. H. 3A) wisely remarks that here "greater precision may be achieved by the use of a formula incorporating the names of the infraspecific taxa than by the use of a collective epithet".

14 × 22. Salix aurita L. × S. herbacea L. = Salix × margarita F. B. White

A low, decumbent or ascending shrub, usually less than 30 cm high; twigs dark, rather glossy, reddish brown, at first thinly pubescent, soon becoming glabrous. Leaves suborbicular, 0.8–3 cm diam., dark green and glabrous above, slightly paler below and pubescent or shortly pilose along the prominent midrib and nerves; leaf margins rather coarsely and bluntly serrate; petiole reddish, up to 1.5 cm long, flanked at the base with 2 well-developed but caducous, ear-shaped, denticulate stipules. Catkins about 1 cm long, appearing with the leaves in April or May, shortly stalked,

---

A: leaves × 2/3; B: stripped twig × 2/3; C: stipule × 3; D: ♂ catkins × 2/3; E: detail × 8; F: ♀ catkins × 2/3; G: detail × 8.

Salix × multinervis 14 × 13

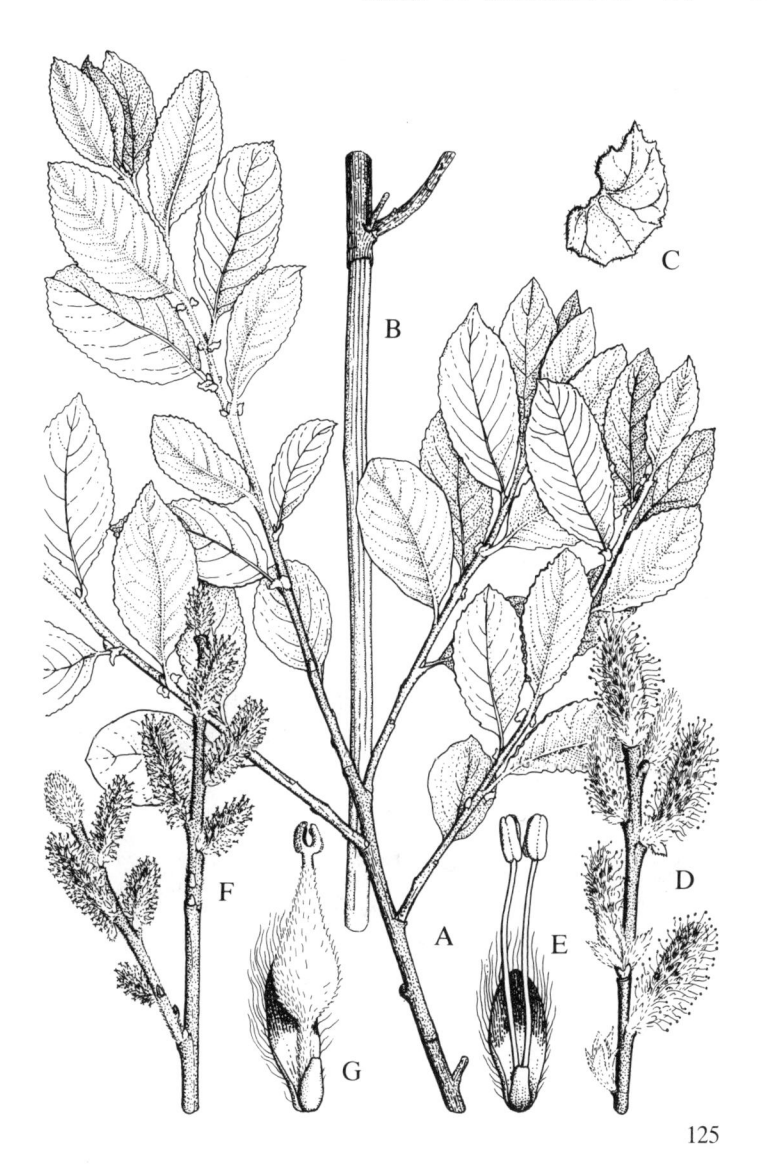

with a few ovate, thinly pilose, foliaceous bracts. Catkin-scales narrowly oblong, obtuse, pale brown, thinly pilose, about 2 mm long, less than 1 mm wide; nectary narrowly oblong; ovary narrowly ovoid, grey-pubescent, about 1.5–2 mm long, 0.8 mm wide, shortly but distinctly pedicellate; style short; stigmas distinctly 2-cleft.

A rare hybrid, figured here to show its affinity with *Salix × grahamii* Borrer ex Baker and *S. × moorei* F. B. White. It was first collected by J. Sadler on Ben Challum, Perthshire, in 1876, and subsequently cultivated in the Royal Botanic Garden, Edinburgh, where F. B. White obtained the material used in drawing up his description of *S. × margarita* (*31* p. 441). Since then the hybrid has been found in several Scottish localities, and has also been reported from Norway.

14 × 22 × 17. Salix aurita L. × S. herbacea L. × S. repens L. = Salix × grahamii Borrer ex Baker

A low, trailing or decumbent shrub, usually less than 30 cm high; twigs dark, purplish-brown, rather glossy, at first thinly pubescent, soon becoming glabrous. Leaves broadly ovate-oblong, 1.5–4 cm long, 1–3 cm wide, at first thinly pubescent on both surfaces, soon becoming glabrous and rather dark, glossy green above, slightly paler below and persistently, but very sparingly, pubescent along the prominent midrib and nerves; apex of leaf obtuse or shortly acute, margins rather coarsely and bluntly crenate-serrate, base rounded; petiole stout, usually less than 8 mm long, reddish, thinly pubescent or subglabrous; stipules up to 8 mm long, ovate, caducous, with glandular-denticulate margins. Catkins cylindrical, rather lax, about 1–1.5 cm long, 0.3–0.4 cm wide, appearing with the leaves in April or May, terminal on short lateral shoots with foliaceous bracts. Catkin-scales oblong-obovate, about 2.5–3.5 mm long, 1.5–2 mm wide, obtuse, pale brown tinged red at the apex, thinly pilose; nectary narrowly oblong, tapering towards apex; ovary flask-shaped with a long slender neck, about 3–3.5 mm long,

---

S. × margarita A: leaves × 2/3; B: ♀ catkins × 2/3; C: detail × 12; *S. × grahamii* D: leaves × 2/3; E: ♀ catkins × 2/3; F: detail × 12; *S. × grahamii* var. *moorei* G: leaves × 2/3; H: ♀ catkins × 2/3; I: detail × 12.

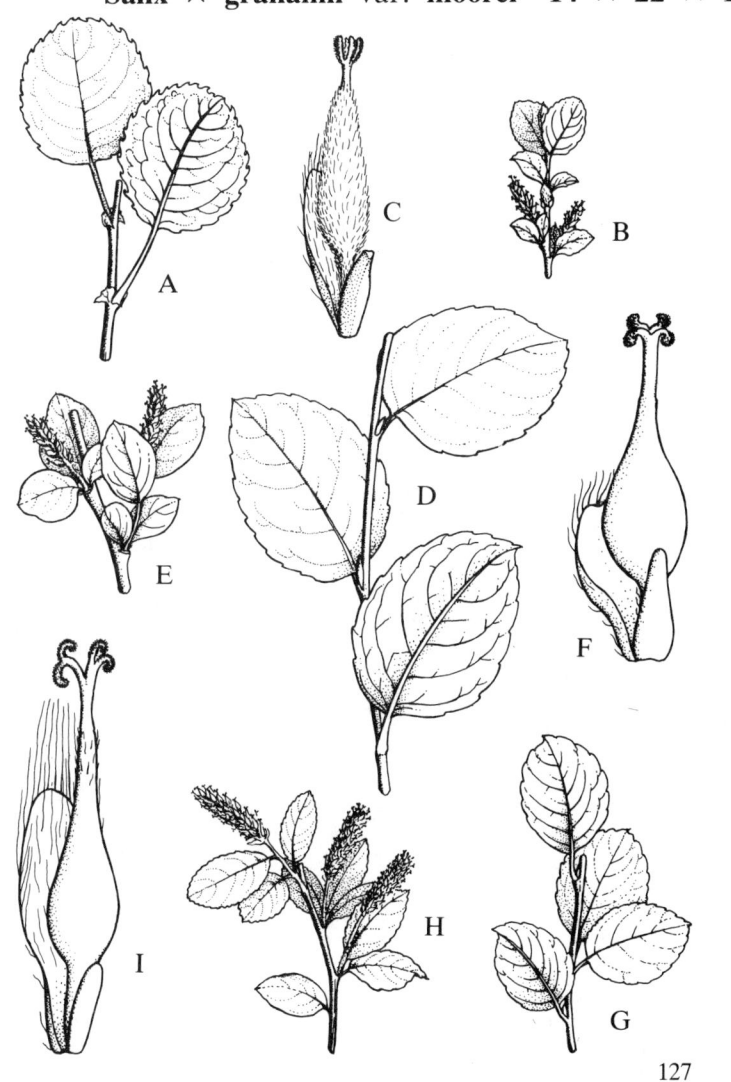

Salix × **margarita**   **14 × 22**
Salix × **grahamii**   **14 × 22 × 17**
Salix × **grahamii** var. **moorei**   **14 × 22 × 17**

1–1.5 mm wide, glabrous, shortly but distinctly pedicellate with a pubescent pedicel; style well developed, reddish; stigmas deeply 2-cleft into linear, recurved lobes.

All records for this mysterious hybrid trace back to a specimen said to have been collected by Professor Robert Graham on Foinaven ("Frouvyn"), Sutherland in 1827 or 1833, and subsequently cultivated in the Royal Botanic Garden, Edinburgh. Although given the manuscript name "Salix Grahami" by William Borrer in May 1856, this was not validly published until 1867 (J. G. Baker in *Journ. Bot.*, 5: 157, t. 66), when an excellent colour plate of the Willow accompanied the description. Baker thought that one of the three sheets in Borrer's herbarium, Kew, had been collected in the wild, but all are marked "Garden", and had probably been growing at Edinburgh for at least twenty three years before being sent to Borrer. A note by H. C. Watson (in the *Compendium* of the *Cybele Britannica*, 576 (1870)) casts some doubt on the origin of *Salix* × *grahamii*: "Professor Graham pointed out this willow to me in the Edinburgh Garden, in the presence of Mr Macnab, assuring me that it was *herbacea* when brought from Sutherland". The inference is obscure, but suggests that *S.* × *grahamii* may have arrived from some other source, but it is more likely that, in its depauperate, original condition, it was misidentified as *S. herbacea*.

Various suggestions have been made as to the parentage of *S.* × *grahamii*: F. B. White thought it was *S. herbacea* × *S. phylicifolia*; E. F. Linton suggested *S. herbacea* × *S. myrsinites*, while the Kew sheets are annotated "*S. arenaria* × *herbacea* × *repens*" by Floderus. All authors agree that *S. herbacea* is one of the parents, and, in my own opinion, Floderus's conjecture is the most likely, though, to account for differences between *S.* × *grahamii* and other material of *S. herbacea* × *S. repens* (*S.* × *cernua* E. F. Linton), I am inclined to the view that it is a hybrid between *S.* × *ambigua* Ehrh. (*S. aurita* × *S. repens*) and *S. herbacea*.

One of the specimens cited by E. F. Linton (*16* p. 82) under *S.* × *grahamii* is certainly incorrect: the sheet in Borrer's herbarium, Kew, labelled "Sow of Athol. J. Ball. From Charles C. Babington" is almost certainly *S. herbacea* × *S. repens*. It is not *S.* × *grahamii*, nor is "*S. herbacea* var. *elliptica* Grev. MS." from "Clova Mountains", though this may be correctly assigned by Linton to *S. herbacea* × *S. myrsinites*. The third specimen mentioned by Linton: "*S. herbacea* L. Rocks above Loch Ceannder, Aug. 1830", has not been examined, though a specimen in herb. Borrer, labelled "Rocks 2 miles above the head of Loch Callader. Aug. 1830. Dr

Greville" is an isotype of *S. macnabiana* Macgillivray (*Edinb. New Phil. Journ.*, 9: 335 (1830)) and is a form of *S. myrsinites*.

S. × *moorei* F. B. White differs from *S.* × *grahamii* only in having rather longer, narrower catkin-scales, and thinly (or occasionally rather densely) pilose ovaries, with glabrous pedicels. Without doubt, it must have the same parentage as *S.* × *grahamii*, of which it can be considered no more than a variety: *S.* × *grahamii* Borrer ex Baker var. *moorei* (F. B. White) Meikle (*Salix* × *moorei* F. B. White in *31* p. 438). It was first found by David Moore "on the top of Muckish mountain, county Donegal" in September 1866 (*Journ. Bot.*, 8: 209 (1870)). Living plants were taken to Glasnevin Botanic Garden, and in due course compared with cultivated material of *S.* × *grahamii*. David Moore (*Journ. Bot.*, 9: 300 (1871)) noted "I was able to get living plants of *S. Grahamii* from Athol [*sic*], which I had planted along with the Muckish plant, and now both are growing freely near to each other, showing, as they do, unmistakably, that they belong to the same species, only differing in some minor points". This conclusion was approved by the Rev J. E. Leefe, the leading British salicologist of the period (*Journ. Bot.* 9: 363 (1871)), but has been challenged by F. B. White (*31*), who thought *S.* × *grahamii* was *S. herbacea* × *S. phylicifolia*, and *S.* × *moorei*, *S. herbacea* × *S. myrsinifolia (nigricans)*. E. F. Linton considered *S.* × *grahamii* to be *S. herbacea* × *S. myrsinites* and referred *S.* × *moorei* to *S. herbacea* × *S. phylicifolia*. But since *S. myrsinites* has not been recorded for Ireland, and since both *S. phylicifolia* and *S. myrsinifolia* are very rare there, and not known to occur in the vicinity of Muckish, I prefer to equate *S.* × *grahamii* and *S.* × *moorei*, and regard both as hybrids between *S. aurita*, *S. herbacea* and *S. repens*.

14×17. Salix aurita L. × S. repens L. Salix × ambigua Ehrh.

Generally a low, sprawling shrub forming loose mats, occasionally erect, and up to 1.5 m high; twigs slender, at first thinly or rather densely pubescent, but later glabrous and dark reddish-brown; underlying wood smooth or sparsely striate. Leaves oblong or obovate, 0.8–4.5 cm long, 0.4–2 cm wide, at first often white with adpressed, silky hairs, glabrescent above with age, and usually dark, dull green, ashy-grey below, and usually remaining rather densely adpressed-sericeous, apex obtuse or acute, sometimes twisted obliquely, margins narrowly recurved, subentire or more or less distinctly undulate-serrate, nervation not very prominent below; petiole very short, usually less than 4 mm long; stipules generally persistent and conspicuous, ear-shaped, up to 5 mm long, 3 mm wide, entire or serrate, sometimes wanting. Catkins appearing before the leaves in April and early May, erect or suberect, sessile or shortly pedunculate, ovoid or very shortly cylindrical, about 8–12 mm long, 4–5 mm wide at anthesis; bracts short, narrowly ovate, densely adpressed-sericeous; catkin-scales oblong or obovate, fuscous or pale reddish-brown, rather densely hairy, 0.8–1 mm long, 0.5–0.8 mm wide, apex rounded or obtuse. Male catkins with 2 free stamens, filaments long, glabrous, much exceeding the catkin-scales; anthers yellow, rather narrowly oblong; nectary oblong, truncate. Female catkins with densely grey-tomentose, shortly pedicellate, narrowly ovoid or flask-shaped ovaries, the pedicel usually shorter than the subtending catkin-scale; style short but usually distinct; stigmas narrowly or broadly oblong, erect, frequently connivent, or spreading, usually entire.

Credit for the discovery of this frequent hybrid must go to William Borrer, who, according to the Rev. W. M. Hind (*Flora of Suffolk,* 320 (1889)) found it, in 1804, at Hopton, Suffolk. No mention was made of the discovery, however, until 1830, when Hooker (*British Flora,* ed. 1, 421) added S. × ambigua (and a number of variants of it) to the British list; by then it was already recorded from stations as far apart as Essex, Suffolk, Isle of Staffa, Perthshire and Angus. Four years later it was figured in *English Botany, Supplement* 2 (t. 2733), and Borrer, who provided the accompanying text, mentions that the German salicologist W. D. J. Koch considered it a hybrid. It seems strange that its hybridity, so obvious to present-day observers, should remain questionable as late as 1834, but the disciples of J. E. Smith remained loyal to their master in refusing to countenance hybrids in the genus. In fact, most populations of *S.* × *ambigua* are convincingly intermediate between *S. aurita* and *S. repens,* and are not difficult to recognize. Occasionally dwarfed or starved plants of *S. aurita* or *S. cinerea* ssp. *oleifolia* are mistaken for the hybrid, but these are quite without the glossy, silky indumentum characteristic of *S.* × *ambigua*. It should

---

A,B,C: leaves × 2/3; D: leaf base × 3; E: ♂ catkins × 2/3; F: detail × 8; G: ♀ catkins × 2/3; H: detail × 8.

Salix × ambigua 14 × 17

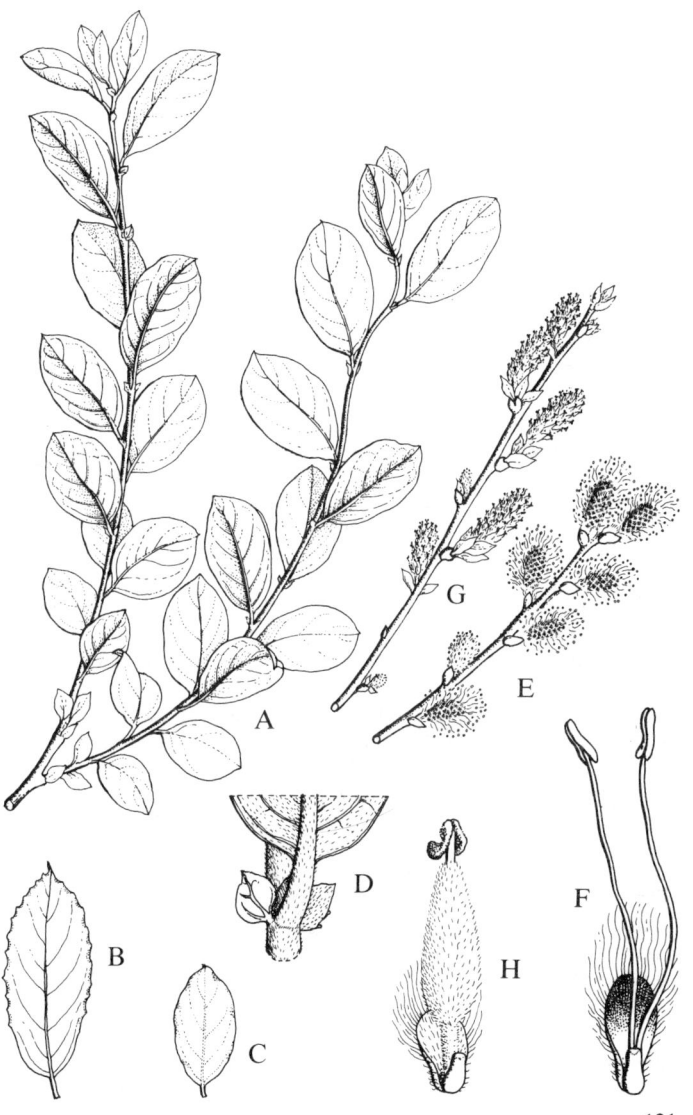

also be noted that the leaves of *S.* × *ambigua*, like those of *S. repens*, tend to blacken in the plant-press.

The hybrid is very widely distributed on acid, heathy ground, occurring wherever both parents are to be found, but, like other hybrids, often uncommon where both parents are equally represented, and, contrary to what one might expect, more frequent where one of the parents is much less abundant than the other. It would appear to be no less common on the Continent than in Britain and Ireland.

14 × 9. Salix aurita L. × S. viminalis L. Salix × fruticosa Doell

Erect, much-branched shrub or small tree, generally less than 5 m high; twigs rather slender, at first densely or thinly grey-pubescent, often (but not always) becoming glabrous and dark reddish-brown with age; underlying wood marked with prominent, scattered *striae*. Leaves lanceolate, often narrowly so, 4–10 cm long, 0.7–2 cm wide, dark dull green and subglabrous or thinly pubescent above, ashy-grey below and softly lanuginose or tomentellous, apex acute or acuminate, base rather narrowly cuneate, margins recurved, subentire or irregularly and bluntly denticulate, generally undulate; nervation often prominent below; petiole short, seldom exceeding 6 mm; stipules conspicuous and persistent, acuminate, with dentate-undulate margins. Catkins appearing before the leaves in April, erect, subsessile, shortly cylindrical, 1–3.5 cm long, about 0.5–0.7 cm wide; bracts lanceolate, 0.5–0.8 cm long, foliaceous, greenish above, densely silvery-hairy below; catkin-scales narrowly oblong, acute or rather bluntly acuminate, pale reddish-brown, rather densely hairy, about 2 mm long, 0.8–1 mm wide. Male catkins with 2 free stamens, filaments glabrous; anthers yellow; nectary narrowly oblong, truncate. Female catkins with densely grey-tomentose, distinctly pedicellate, narrowly ovoid or flask-shaped ovaries 2–2.5 mm long, about 1 mm wide, pedicel about 0.7 mm long, usually longer than the narrowly oblong nectary; style distinct, up to 0.5 mm long; stigmas linear or narrowly oblong, about 0.8 mm long, spreading, usually entire.

A: leaves × 2/3; B: stripped twig × 2/3; C: stipule × 3; D: ♂ catkins × 2/3; E: detail × 8; F: ♀ catkins × 2/3; G: detail × 8.

Salix × fruticosa   14 × 9

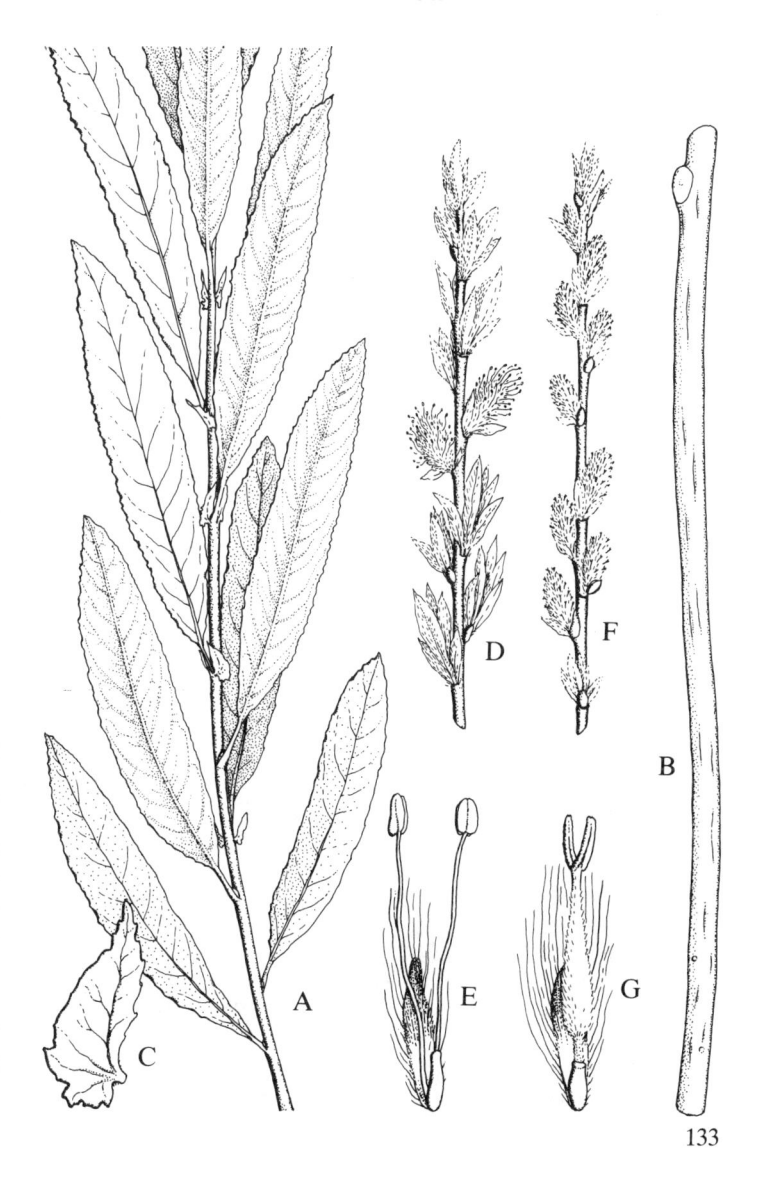

Not a common hybrid in our area, but widely scattered in England, Scotland and Ireland, and perhaps more frequent than one would expect considering the very different flowering periods of the two parents. *S. × fruticosa* is known to have been cultivated by basket-makers, and its wide distribution may, in part, be the result of former cultivation, since many of the specimens are so uniform as to suggest derivation from a single clone. It is, however, quite certainly spontaneous in Ireland and Scotland, where hybrid populations appear to be more varied than in England.

It is readily identifiable by its undulate-margined leaves and conspicuous, persistent stipules. The nervation and indumentum of the leaf-undersurface resembles that of *S. caprea × S. viminalis,* but in other respects the two hybrids are easily distinguished.

As with other Willow hybrids, female plants are much more common than males, and the male flowers illustrated have been taken from Swedish specimens, since no British or Irish example could be found. *Salix × fruticosa* is frequent in northern Europe, but Continental specimens mostly reflect the slight, but uniform differences between the *S. aurita* of our area and that of the Continent.

The plant figured as *Salix ferruginea* in *English Botany Supplement,* t. 2665, is, to judge from specimens in Borrer's herbarium, Kew, definitely *S. aurita* L. × *S. viminalis* L. It is not, however, *Salix ferruginea* G. Anders. ex Forbes (*9* p. 255, t. 128) as stated; the latter (represented in Borrer's herbarium) is *S. cinerea* L. ssp. *oleifolia* Macreight × *S. viminalis* L.

15. Salix myrsinifolia Salisb.

Dark-leaved Willow

A very variable shrub or small tree, sometimes low and sprawling, and less than 1 m high, but often forming a robust, open bush, more than 3 m high, sometimes with a distinct trunk; bark dark greyish, lightly fissured; twigs dull brown or greenish, at first densely pubescent, and usually not becoming quite glabrous until more than

---

A: leaves × 2/3; B: stipule inside × 6; C: ♂ catkins × 2/3; D: detail × 8; E: ♀ catkins × 2/3; F,G: details × 8.

Salix myrsinifolia 15

a year old, wood of peeled twigs sometimes with a few distinct *striae*; buds generally blunt and pubescent; leaves obovate, elliptic or oblong, 2–6.5 cm long, 1.5–3.5 m wide, rather thin and papery, sublustrous dark green above, glaucous below (the glaucosity often vanishing towards the leaf-apex, or leaves sometimes green on both surfaces in shade-grown specimens), turning blackish when dried, at first thinly pubescent with straight subadpressed hairs above and below, becoming subglabrous with age, but the midrib generally remaining pubescent below even at maturity; nervation prominent or obscure; apex usually shortly acute, base cuneate or rounded; margin not, or very narrowly, recurved, distinctly but rather irregularly serrate, rarely subentire; petiole generally less than 1 cm long; stipules variable, frequently well developed and persistent, but sometimes small and caducous or apparently absent, ear-shaped, normally pubescent with a few, small, scattered, sessile glands. Catkins appearing with the leaves in April and May, suberect, usually terminal on short lateral shoots, occasionally subsessile, cylindrical, 1.5–4 cm long, 1–1.5 cm wide; bracts 2–4, often more than 1 cm long, 0.5–1 cm wide, thinly or densely adpressed-pubescent; catkin-scales shortly oblong or suborbicular, 1–2.5 mm long, 0.8–1.5 mm wide, fuscous, rather thinly hairy. Male catkins with 2 free stamens; filaments glabrous throughout or hairy in the lower half, up to 1 cm long; anthers yellow, oblong, about 1 mm long, 0.7 mm wide, nectary shortly oblong, about 0.7 mm long, truncate. Female catkins often longer, narrower and less compact than the male, ovary flask-shaped, 2.5–3 mm long, 0.8 mm wide, glabrous or pubescent, or with pubescent zones, usually tapering to a slender neck; pedicels pubescent, about 1 mm long; style usually distinct; stigmas suberect or spreading, about 0.5 mm long, sometimes entire and oblong, but commonly cleft into four linear lobes. Capsules up to 7 mm long when mature, usually with a distinctly lengthened pedicel,

This species and *Salix phylicifolia* L. are unquestionably the two most problematic willows in our area. So elusive are the distinctions between them in this area that Buchanan White (*31*) was driven to regarding them as "subspecies or major varieties" of a single species. E. F. Linton (*16*), following Enander and other authors, tried to restrict the name *S. myrsinifolia* (or rather *S. andersoniana* Sm.) to plants with glabrous ovaries, pilose filaments and leaves serrate to the apex, referring to *S. phylicifolia* plants with hairy ovaries, glabrous filaments and entire-tipped leaves, while specimens exhibiting a mixture of these characters were determined as *S.*

myrsinifolia (andersoniana) × *S. phylicifolia*. This is a neat solution to the problem, but not a very convincing one, particularly as the resultant "hybrids" may come from areas where only one of the putative parents is known to occur. A final answer has yet to be found to the *myrsinifolia-phylicifolia* group in Britain, meantime the two species can *generally* be identified as follows:

S. myrsinifolia	S. phylicifolia
Twigs dull brown or greenish, usually pubescent.	Twigs glossy brown, usually glabrous.
Leaves thin, papery, sublustrous dull green, blackening when dried.	Leaves rather rigid and coriaceous, bright shining green above, not blackening when dried.
Stipules usually developed and persistent.	Stipules commonly very small, caducous or wanting.
Ovaries commonly glabrous.	Ovaries commonly hairy.

The name *S. nigricans* Sm. (1802), familiar to generations of British and Irish botanists, must, most regrettably, be replaced by *S. myrsinifolia* Salisb. (1796), since, under current rules of nomenclature, this would appear to be the earliest validly published name for the species.

Although always rather local, *S. myrsinifolia* has an extensive distribution in England and Scotland north of central Lancashire and Yorkshire, usually growing by river banks and lake shores at altitudes below 600 m (2000 ft). It is rare, as a native plant, in Ireland, and has been largely lost sight of there since 1833, when it was recorded from several localities in counties Antrim and Derry by David Moore (see *Irish Nat. Journ.*, 10: 38–41 (1950)). The recent re-finding of specimens, identified as *S. cinerea* L. ssp. *oleifolia* Macreight × *S. myrsinifolia* Salisb., and collected in 1931 by A. W. Stelfox at Glendun, Co. Antrim (one of Moore's localities for *S. myrsinifolia*), suggests that the species still survives in this area (see *Irish Nat. Journ.*, 19: 77–79 (1977)). Records for Surrey, Warwickshire, Kildare, Westmeath and Cavan are based on introduced plants; the reasons for introduction are obscure, since *S. myrsinifolia* is neither ornamental nor useful. Recent records from Suffolk, Cambridge and Lincoln are, however, apparently based on indigenous populations, and suggest that *S. myrsinifolia* may extend further south than has hitherto been supposed; it also means that the original material of *S. nigricans* may truly have been collected by James Crowe at Wormegay ("Wrongay") Fen, Norfolk, as stated by Smith (*Fl. Britannica*, 1047 (1804)), though there are no recent records for the species from this area.

Outside our area, *S. myrsinifolia* has a distribution similar to, but more extensive than that of *S. phylicifolia*, ranging from Ireland eastwards to Siberia, with a detached area in the mountainous regions of central Europe.

16. Salix phylicifolia L.

Tea-leaved Willow

A robust, much-branched shrub or small tree, generally about 2–3 m high, but occasionally up to 5 m; bark greyish, lightly fissured; twigs usually rich reddish-brown, at first sometimes thinly and shortly pubescent, but soon glabrous and rather lustrous; wood of peeled twigs generally without striae; leaves rather rigid and coriaceous, bright shining green above, glaucous below, not turning black on drying, oblong, ovate, elliptic or suborbicular in outline, 2–6 cm long, 1–5 cm wide, at first sometimes very thinly pubescent, soon becoming glabrous on both sides; nervation generally not very prominent, apex acute or obtuse, base cuneate or rounded, margins not, or very narrowly, recurved, entire, subentire, bluntly serrate or occasionally sharply and distinctly serrate; petiole short, rigid, seldom exceeding 1 cm in length; stipules wanting or minute and caducous, rarely well-developed and broadly ear-shaped with toothed margins. Catkins appearing with the leaves in April and May, usually terminal on short lateral shoots, sometimes subsessile, cylindrical, 1.5–4 cm long, 1–1.5 cm wide; bracts 2–4, foliaceous, usually less than 1 cm long and 0.8 cm wide, thinly or densely sericeous-pubescent on the lower surface, usually glabrous above; catkin-scales oblong, obovate or suborbicular, 1–2.5 mm long, 0.8–1.5 mm wide, fuscous, thinly or densely hairy. Male catkins with 2 free stamens, filaments glabrous, up to 1 cm long, anthers yellow, oblong, about 0.6 mm long, 0.4 mm wide, nectary shortly oblong, about 0.7 mm long, truncate. Female catkins about as long as the male, usually rather dense, ovary flask-shaped, about 2–3

A: leaves × 2/3; B: ♂ catkins × 2/3; C: detail × 8; D: ♀ catkins × 2/3; E,F: details × 8.

Salix phylicifolia 16

mm long, 0.8–1 mm wide, densely ashy-pubescent (or sometimes glabrous or subglabrous) usually tapering to a slender neck; pedicel very short or almost absent at anthesis; style usually distinct, about 1 mm long; stigmas about 0.5 mm long, usually spreading and cleft into 4 linear lobes. Capsule up to 7 mm long at maturity, usually remaining subsessile or very shortly pedicellate.

The differences between this and *Salix myrsinifolia* Salisb. (*S. nigricans* Sm.) are summarized under the latter species. On the whole it can be said that *S. phylicifolia* is the less variable of the two, and that it is normally recognizable by its shining twigs and bright green, glabrous, rather coriaceous leaves. In habit and leaf-shape it is excessively polymorphic, and distinctions based on either character are quite worthless. Most British and Irish specimens have densely pubescent ovaries and capsules, but the glabrous var. *lejocarpa* Anderss. is occasionally collected.

S. phylicifolia has in our area a distribution largely overlapping with that of *S. myrsinifolia (nigricans)*; it is locally abundant on moist rocky ground, commonly on carboniferous limestone, from Lancashire and Yorkshire north to the Orkneys, and has been recorded from near sea-level up to 670 m (2200 ft) in the Highlands. It is very rare in Ireland, being apparently confined to the limestone of the Ben Bulben range, in counties Leitrim and Sligo. The Irish plant has been distinguished as *S. hibernica* Rechinger f., but the differences, based upon shape and nervation of the leaves, are unsatisfactory, nor is the Irish material homogeneous. Abroad, *S. phylicifolia* has a wide distribution in northern Europe, from Iceland and Scandinavia to the U.S.S.R., it has also a few isolated stations in central Europe, in S.E. Germany, Austria and Czechoslovakia.

17. Salix repens L.

Creeping Willow

A prostrate, decumbent, ascending or erect shrub or subshrub, up to 1.5 m (or, it is reported, as much as 2 m) high; twigs slender or

---

A,B,C,D,E: leaves × 2/3; F: stipule × 8; G: ♂ catkins × 2/3; H: detail × 12; I: ♀ catkins × 2/3; J: detail × 12.

Salix repens and variants 17

rather robust, fuscous-grey, or reddish, or yellowish-brown, glabrous, pubescent or densely adpressed-sericeous; leaves 1–3.5 cm long, 0.4–2.5 cm wide, lanceolate, oblong or ovate-oblong, glabrous, pubescent or silvery-sericeous on both sides, or glabrous and bright green above and pubescent or sericeous below, drying black, margins generally rather sharply recurved, entire or obscurely glandular-serrulate, apex blunt or shortly mucronate or acute, sometimes twisted obliquely, base cuneate or rounded, nervation obscure; petiole very short, usually less than 4 mm long; stipules commonly wanting, sometimes well developed and persistent, lanceolate or narrowly oblong, acute, 2–3 mm long, 1–1.5 mm wide, entire or minutely and remotely toothed. Catkins appearing with, or, more usually, before the leaves in April and May, erect or suberect, sessile or subsessile, ovoid or obovoid or shortly cylindrical, 1–2.5 cm long, 0.4–0.8 cm wide, the females usually rather smaller than the males; bracts small and inconspicuous, sometimes well developed and foliaceous, commonly pubescent or sericeous, generally less than 5 mm long and 2 mm wide; catkin-scales obovate-lingulate, about 2 mm long, pale or stained reddish-brown towards the apex, subglabrous or thinly pilose, sometimes rather densely sericeous-villose. Male catkins with 2 free stamens, filaments glabrous or occasionally hairy at the base, 5–8 mm long; anthers yellow, oblong, about 0.5 mm long, 0.4 mm wide; nectary oblong, about 0.6 mm long, truncate at apex. Female catkins with subsessile or shortly pedicellate, narrowly flask-shaped, glabrous, pubescent or sericeous-tomentose ovaries, 2–2.5 mm long, 1 mm wide; pedicel pubescent or sericeous, rarely glabrous, about 0.5 mm long; style about 0.4 mm long, glabrous, usually conspicuous; stigmas 2, shortly bifid or entire. Capsules up to 7 mm long and 2.5 mm wide at maturity.

A polymorphic plant, represented in Britain and Ireland by a whole spectrum of variants ranging from slender, prostrate, glabrous subshrubs at one extreme, to relatively robust, erect or ascending, sericeous-leaved shrubs at the other, sometimes forming distinct local populations, but more often intergrading. Many attempts have been made to devise a workable analysis, but none wholly successful, and the synonymy of *Salix repens* is a memorial to these frustrated endeavours. The most successful is probably that of

---

A: leaves × 2/3; B: stipule × 4; C: ♂ catkins × 2/3; D: detail × 8; E: ♀ catkins × 2/3; F: detail × 8.

Salix repens var. argentea 17

C. E. Moss (*18* p. 49–51). Here aggregate *S. repens* is divided into three varieties, which are named: 1. *S. repens* L. var. *ericetorum* Wimm. et Grab., the common procumbent or decumbent variety of heaths and moorlands, often with small, subglabrous leaves and ovaries. 2. *S. repens* var. *fusca* Wimm. et Grab., an erect or ascending plant, up to 1.5 m or more in height with the ovaries and lower surface of the leaves often densely sericeous-hairy. This variant is locally common, and tolerably distinct, in the fenlands of East Anglia. 3. *S. repens* L. var. *argentea* (Sm.) Wimm. et Grab., normally, but not always, a plant of maritime dunes, with robust, ascending branches, pubescent twigs, and large, broad, blunt leaves densely clothed with silky, silvery hairs on both surfaces, or on the lower surface.

The first variant should be renamed *S. repens* L. var. *repens,* unless one accepts Floderus's hypothesis (see Fraser in *12* p. 370) that true *S. repens* var. *repens* has glabrous pedicels and ovaries, and that most, or perhaps all of the British material so named consists of hybrids between *S. repens* L. and *S. arenaria* L. (= *S. repens* L. var. *argentea* (Sm.) Wimm. et Grab.). This hypothesis would certainly explain the excessive polymorphism of *S. repens* aggr., but it is equally possible that *S. repens* L. sec. Floderus and *S. arenaria* L. sec. Floderus are no more than extreme variants of one species. *S. repens* L. var. *argentea* (Sm.) Wimm. et Grab. (*S. arenaria* L. sec. Floderus) is a common plant of dune slacks all round Britain and Ireland, and many occasionally be found inland on sandy slopes and banks. Although sometimes distinct and uniform, it frequently merges with other variants.

The second of Moss's varieties, *S. repens* L. var. *fusca* Wimm. et Grab., a characteristic fenland plant, is, in its typical condition, a remarkable variant, with numerous erect branches and twigs. While it is possibly the *Salix fusca* of Smith and Forbes, or the *S. repens* L. var. *fusca* of Wimmer and Grabowski, it scarcely agrees with the original *S. fusca* of Linnaeus, which is described (*Fl. Suecica*) as a creeping plant with very small leaves. Moreover, as with *S. repens* var. *argentea,* it is by no means uniform, some populations including plants tending towards *S. repens* var. *repens,* others approaching *S. repens* var. *argentea.* Such intermediates include *S. adscendens, S. parvifolia, S. incubacea, S. foetida,* and other segregates named and described by Smith, Borrer and others in the early years of the last century.

S. repens, in the wide sense, occurs locally almost throughout Britain and Ireland, but is generally rare in districts where heaths and acid, sandy or peaty ground are wanting. It has an extensive

European distribution north of the Mediterranean, where it intergrades, towards the eastern limits of its range, with the closely allied *S. rosmarinifolia* L., a narrow-leaved plant which has been erroneously recorded from the British Isles.

17 ×9. Salix repens L. × S. viminalis L. = Salix × friesiana Anderss.

Slender, erect or sprawling shrubs, 0.5–2 m high; young shoots at first densely tomentellous, remaining rather densely pubescent through their first year, but becoming dark brown and glabrous with age. Leaves rather crowded, lanceolate, 4–7 cm long, 0.5–1.5 cm wide, dull green and thinly pubescent or subglabrous above, thinly or densely sericeous-pubescent or tomentose below, apex acuminate, margins narrowly recurved, entire or subentire, sometimes undulate. Petiole very short, generally less than 5 mm long. Stipules lanceolate, 4–5 mm long, 1–1.5 mm wide, remotely glandular-serrulate, generally caducous. Catkins appearing before the leaves in April, sessile or subsessile, oblong-ovoid, the males 1–1.5 cm long, 0.7 cm wide, the females generally rather longer in proportion to their width; bracts few, lanceolate, 5–7 mm long, 2–3 mm wide, sericeous below; catkin-scales densely pilose, ovate-oblong, 2–3 mm long, 1.5 mm wide, brownish all over, with a darker acute, subacute or rounded apex; nectary oblong, truncate, about 0.3 mm long; male flowers with 2, free, glabrous filaments 3–4 mm long; anthers yellow, oblong, about 0.8 mm long; female flowers with narrowly flask-shaped, densely silvery-sericeous ovaries 2–3 mm long, 1 mm wide; style short, about 0.8 mm long; stigmas linear, undivided or 2-cleft, about 0.5–0.8 mm long. Capsules 3–5 mm long, 2–2.5 mm wide, greyish pubescent; probably sterile.

A very rare hybrid, first collected in Britain by E. S. Marshall, who found a single bush on river gravel near Brora in E. Sutherland, on August 9th, 1897 (*E. S. Marshall* 1928). Quite recently the hybrid has turned up, in much greater abundance, on sandhills near Southport, S. Lancs. (VC 59). The Southport specimens much resemble a deliberate hybrid, manufactured by E. F. Linton at the end of the 19th century, and distributed as no. 98 of his *Set of British Willows*. The dense sericeous indumentum on the lower surface of the leaves suggests *S. repens* L. var. *argentea* (Sm.) Wimm. et Grab. rather than typical *S. repens* as one of the parents.

Abroad, *S. repens* × *S. viminalis* is on record from Denmark, Norway, Sweden, Russia, Austria, Germany and Yugoslavia, but it is nowhere common, and may have been planted in some of its localities.

18. Salix lapponum L.

Downy Willow

A low, much-branched shrub (15–)20–100(–150) cm high; twigs rather rigid with prominent bud-scars, dark reddish-brown, at first pubescent or lanuginose, soon becoming glabrous and rather glossy; buds dark brown, glabrous or pubescent; leaves usually lanceolate or narrowly obovate, sometimes ovate or oblong, 1.5–7 cm long, 1–2.5 cm wide, dull grey-green and thinly or rather densely adpressed-villose above, pale grey below and often densely tomentose or lanuginose, rarely subglabrous on both surfaces; margins entire or subentire, sometimes a little undulate, narrowly recurved or flattish, apex usually acute or acuminate, sometimes twisted obliquely, base generally cuneate; petiole short, occasionally up to 1 cm long, but usually less than 5 mm; stipules commonly wanting, small, narrowly ear-shaped, acuminate, often falcately curved, entire or toothed, about 3 mm long, 1.5 mm wide. Catkins appearing with or a little before the leaves in May and June, erect or suberect, sessile or subsessile, shortly cylindrical, 2–4 cm long, 1–1.5 cm wide, the females strongly accrescent, often as much as 7 cm long in fruit; bracts small, inconspicuous, narrowly acute, about 5 mm long, 2.5 mm wide, densely sericeous-lanuginose; catkin-scales narrowly or broadly elliptic or ovate, 2–4 mm long, 1–1.5 mm wide, acute or obtuse, dark fuscous-brown, sometimes paler towards base, sericeous-villose. Male catkins with 2 free stamens, filaments glabrous, 4–6 mm long, anthers yellowish or stained purplish or reddish, shortly oblong, about 0.8 mm long, 0.6 mm

---

A: leaves × 2/3; B: leaf base and stipules × 3; C: ♂ catkins × 2/3; D: detail × 8; E: ♀ catkins × 2/3; F: detail × 8.

Salix × friesiana 17 × 9

wide, nectary narrowly oblong, about 0.5 mm long, truncate at apex. Female catkins with sessile or subsessile, narrowly flask-shaped, densely whitish-tomentose ovaries, about 3–4 mm long, 1–1.5 mm wide, a little longer than the catkin-scales; style very slender and conspicuous, often 1–1.5 mm long; stigmas divided to the base into 4 filiform arms up to 1.5 mm long. Capsules narrowly flask-shaped, up to 8 mm long and 3 mm wide at maturity.

One of the easiest Willows to recognize, varying only slightly in the shape and indumentum of its leaves. Smith's *Salix stuartiana* (*English Botany*, 36: t. 2586 (1814)) from Breadalbane, is one of the more hirsute variants (though this does not come out clearly in Sowerby's illustration); his *S. arenaria* (*English Botany*, 26: t. 1809 (1808)) is typical *S. lapponum* L. The record for *S. helvetica* Vill. "from the Highlands of Scotland" (figured—as *S. glauca*—in *English Botany*, 26: t. 1810 (1808)) is unquestionably erroneous. *Salix helvetica*, a distinct species with discolorous, contrasting leaf-surfaces (the upper surface green and lustrous, the lower white-tomentose) and short, sparsely tomentose ovaries, replaces *S. lapponum* in the central European Alps and is not likely to occur in Britain. The error probably originated in Mr Crowe's garden in Norfolk, whence material of *S. helvetica*, said to be Scottish in origin, but almost certainly Swiss, was distributed to Dawson Turner and other botanists of the period.

S. lapponum is locally common on cliffs and rocky mountainsides north of the lowlands of Scotland, generally at altitudes between 200 and 900 m (700 and 3000 ft); otherwise it is known in our area only from Dumfriesshire and Helvellyn (Westmorland). It has a wide distribution in northern Europe, eastwards to the Altai and western Siberia, and is found as far south as the Pyrenees and Bulgaria.

A,B,C,D,E: leaves × 2/3; F: ♂ catkins × 2/3; G: detail × 8; H: ♀ catkins × 2/3; I: detail × 8.

Salix lapponum 18

19. Salix lanata L.

Woolly Willow

A low-growing, gnarled, much-branched bush, usually less than 1 m high; twigs rather stout and rigid, with prominent leaf-scars, at first thinly lanuginose, but soon becoming glabrous and rich sublustrous brown; buds dark reddish-brown, hairy at first, becoming glabrous with age, the catkin-buds conspicuously large and swollen, more than 1 cm long and almost as wide, with an acute or mucronate apex; leaves rather variable, often broadly ovate or suborbicular, 3.5–7 cm long, 3–6.5 cm wide, but sometimes obovate or elliptic, lamina generally floccose, or thinly lanuginose or cobwebby at first, often becoming subglabrous at maturity, dark or grey-green above, distinctly glaucous below, usually with rather conspicuous, finely reticulate nervation, margins entire or subentire, rarely with a few irregular teeth, flat or very narrowly recurved, apex obtuse or acute, frequently with a small obliquely deflected mucro, base broadly cuneate, rounded, or often cordate; petiole short and stout, usually less than 1 cm long, subacute above, villose or glabrescent; stipules large, conspicuous, often persistent, broadly ovate or obovate, frequently 1 cm long and 0.6 cm wide (exceptionally up to 2 cm long and 1.8 cm wide), entire, glaucous below. Catkins appearing before the leaves from late May to July, erect, sessile, dense, broadly cylindrical, at first 2.5–3.5 cm long, 1.3–1.5 cm wide, the female lengthening considerably (sometimes to 7 cm) in fruit; bracts small, inconspicuous, densely villose, those of the female catkins enlarging and becoming foliaceous with age; catkin-scales ovate or elliptic, 2–3 mm long, 1–2 mm wide, dark fuscous-brown, densely clothed with soft, silky, golden-yellow (rarely silver-grey) hair. Male catkins with 2 free stamens, filaments glabrous, up to 9 mm long, anthers yellow, oblong, about 1 mm long, 0.5 mm wide; nectary oblong or oblong-urceolate, about 0.8 mm long, 0.5 mm wide, truncate at apex. Female catkins with sessile or subsessile, narrowly flask-shaped, glabrous ovaries about 3 mm long, 1 mm wide, hidden under the dense indumentum of the catkin-scales; style distinct, often 1 mm long, sometimes bifid towards apex; stigmas narrowly oblong or linear, 0.6–0.8 mm long, at first

---

A: leaves × 2/3; B: ♂ catkins × 2/3; C: detail × 12; D: ♀ catkins × 2/3; E: detail × 12.

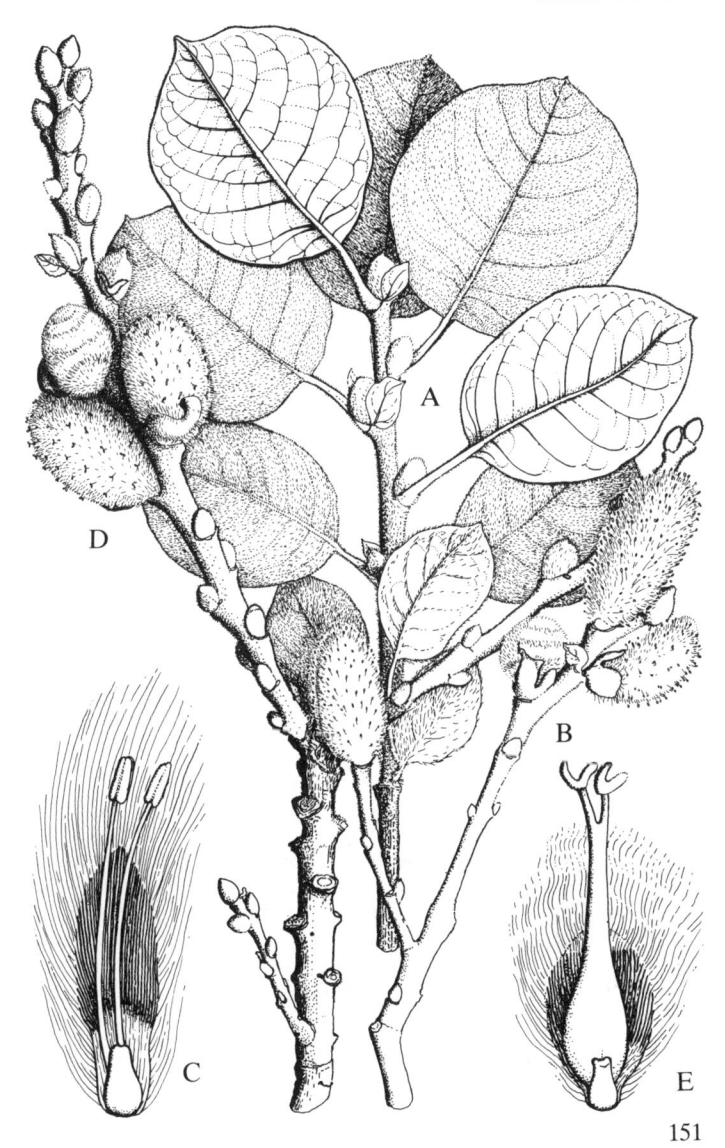

Salix lanata

undivided, becoming deeply bifid with age. Capsules narrowly flask-shaped, about 6 mm long, 2 mm wide at base, very shortly pedicellate.

A subarctic species, distributed from Iceland and the Faeroes through northern Scandinavia, Finland and Russia to eastern Siberia, with close allies (sometimes ranked as subspecies) in North America. It is rare in our area, surviving in a few localities in the highlands of Perthshire, Angus and Aberdeen, on rocky mountainsides at altitudes between 600 and 900 m (2000 and 3000 ft).

Though the leaves vary in shape and indumentum, *Salix lanata* is so distinctive in other respects as to have attracted unusually few synonyms, and none of these in current use. The first British record must be attributed to George Don, who found *Salix lanata* in Glen Callater, S. Aberdeen, in 1812. Smith (*27* p. 205) erroneously claims Thomas Drummond as the discoverer, and evidently knew the species only from "rocks amongst the Clova mountains", where Drummond found it. Loudon (*17* p. 1594) and Druce (*Notes Roy. Bot. Gard. Edinb.*, 12: 131 (1904)) both draw attention to the inaccuracy, which was also corrected by David Don (the son of George Don) in a footnote added by J. de C. Sowerby under *Carex vahlii* in *English Botany, Supplement 1*, t. 2666 (1831).

20. Salix arbuscula L.

Mountain Willow

A low shrub, generally less than 70 cm high with spreading or ascending, repeatedly divided branches; twigs at first sparsely and shortly pubescent, soon becoming glabrous and dark, sublustrous, reddish-brown; leaves generally ovate, less often oblong or elliptic, 1.5–3(–5) cm long, 1–1.5(–3) cm wide, glabrous and bright lustrous green above, densely adpressed-pubescent below at first, but soon becoming glabrous and glaucous or glaucescent, with rather obscure venation; apex of leaf generally acute, margins narrowly recurved, with numerous, small, blunt, gland-tipped teeth, base broadly or

---

A,B,C: leaves × 2/3; D: leaf base × 4; E: ♂ catkins × 2/3; F: detail × 12; G: ♀ catkins × 2/3; H: detail × 12.

Salix arbuscula 20

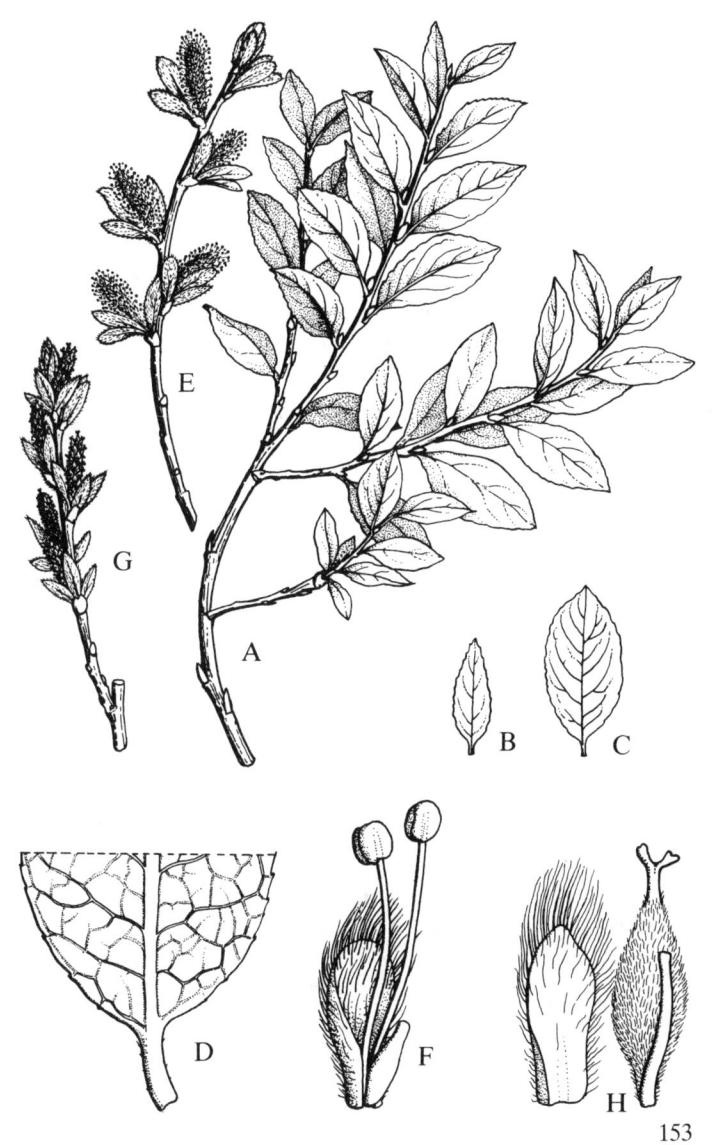

narrowly cuneate; petioles short, usually less than 5 mm long (rarely up to 8 mm); stipules usually wanting, or early caducous, bluntly ovate or ear-shaped, 1.5–2 mm long, 1–1.5 mm wide, with entire or minutely serrulate margins. Catkins appearing with the leaves in late May or June, numerous and erect or suberect on short lateral shoots, narrowly cylindrical, 10–20 mm long, 4–5 mm wide, bracts foliaceous, 0.8–2 cm long, 0.4–0.8 cm wide, glabrous or subglabrous above, at first adpressed-sericeous below, but soon glabrescent; catkin-scales broadly and bluntly ovate or obovate, about 1.3 mm long and almost as wide, brownish or tinged dark purple, densely hairy. Male catkins with 2 free stamens, filaments glabrous, 3–4 mm long, anthers dark reddish or purple, suborbicular, less than 0.5 mm diam., nectary narrowly oblong or elongate flask-shaped, up to 1 mm long, less than 0.5 mm wide, apex blunt or truncate. Female catkins often longer than male, but with proportionately smaller catkin-scales; ovaries flask-shaped, sessile, about 1.5 mm long, 0.8–1 mm wide, densely grey-pilose; style distinct, about 0.5 mm long; stigmas narrowly oblong, rather less than 0.5 mm long, notched or deeply 2-cleft. Capsules broadly flask-shaped, about 3 mm long, 2 mm wide, pubescent even when fully ripe.

First recorded by Lightfoot (1777) from the mountains of Breadalbane, but misidentified as *Salix myrsinites* L. Smith realized that Lightfoot's identification was erroneous, but went on to compound the error by applying the name *S. arbuscula* to a form of *S. repens* L. (*English Bot.* t. 1366) and by giving four different names, *S. prunifolia* Sm., *S. venulosa* Sm., *S. carinata* Sm. and *S. vaccinifolia* Walker, to material which would nowadays be referred to *S. arbuscula* without hesitation. No one has attempted to resurrect Smith's (or Walker's) names, for the supposed distinctions in habit and foliage are too slight to excite the attention even of the most critical salicologist. On the whole, *S. arbuscula* is not very variable, and its small, neat, glossy leaves and small catkins are easily recognized in the field.

S. arbuscula is locally abundant on damp, rocky mountain slopes and ledges in the Breadalbane area of mid-Perthshire, and in adjacent parts of Argyll, but apart from one recent record from Peebles, is virtually confined to this area, though formerly also in Forfar and S. Aberdeen. Records from Dumfries, Inverness and the Orkneys call for re-examination. In Britain the Willow is rarely found below 600 m (2000 ft), and has been found at 820 m (2700 ft) in Breadalbane.

Abroad *S. arbuscula* has a limited distribution, occurring from north Scandinavia eastwards to Siberia. Records from central Europe are misidentifications, mostly of the allied *S. foetida* Schleicher and *S. waldsteiniana* Willd.

As an additional comment, it may be worth noting that what appeared to be an extremely old, gnarled and moribund "trunk" of *S. arbuscula* proved, on examination, to be just twenty-five years of age, so that individual plants of the Willow may not be long-lived.

21. Salix myrsinites L.

Whortle-leaved Willow

A low, spreading or decumbent bush, generally less than 40 cm high, with long, rooting, underground branches; twigs at first very thinly pilose, soon becoming glabrous and dark, lustrous reddish-brown; leaves often persisting in the withered condition, very variable in shape and size, oblong, ovate, obovate or rarely lanceolate, 1.5–7 cm long, 0.5–2.5(–3) cm wide, rich, shining green above and below, at first thinly clothed with long hairs, especially on the undersurface, but soon glabrous with prominently reticulate venation; margins regularly and rather closely glandular-serrulate; apex acute or obtuse, base rounded or cuneate; petiole rather short and stout, usually less than 1 cm long, channelled above; stipules ovate or oblong, up to 5 mm long, 4 mm wide with serrulate margins, generally prominent and rather persistent, but sometimes wanting or caducous. Catkins appearing with the leaves in late May and June, rather few and remote, erect, terminal on short lateral shoots, peduncle usually well developed, up to 1.8 cm long, thinly villose; bracts foliaceous, lanceolate to suborbicular, 1–2 cm long, 0.5–1.5 cm wide, thinly villose or glabrescent with minutely serrulate margins; catkin-scales uniformly fuscous or dark reddish-purple, oblong or obovate, 1.5–2 mm long, about 1 mm wide, villose with long silky hairs, apex acute, obtuse or rounded. Male catkins with 2 free stamens; filaments purplish, glabrous or sometimes thinly pilose near the base, 5–8 mm long; anthers purple or reddish before dehiscence, suborbicular, 0.4 mm diam.; nectary shortly oblong, less than 0.5 mm long, apex truncate. Female catkins usually larger than male, often conspicuously large and accrescent after anthesis, cylindrical, 3–5 cm long, 1.2–2 cm wide (at anthesis), up to 7 cm long in fruit; ovary flask-shaped, 4–5 mm long,

1.5–2 mm wide, at first thinly or rather densely covered with thick, crinkled, iridescent hairs, but sometimes glabrescent with age; style usually distinct, up to 2.5 mm long, rather short; stigmas suberect or spreading, up to 1 mm long, oblong and subentire, or more usually deeply 2-cleft; nectary similar to that of the male flower. Capsule commonly 7 mm long, 2 mm wide, often glabrescent, but usually with a few, persistent, iridescent hairs.

An easy Willow to identify and not very variable: the lustrous green undersurface of the leaves, and the disproportionately large female catkins are distinctive, and, where there are doubts, the characteristic iridescent hairs on the ovary (more conspicuous in dried than in living specimens) will not only resolve them for the species, but also serve as a useful pointer to *Salix myrsinites* hybrids.

S. myrsinites has a wide but scattered distribution in Scotland, from Argyll and Perthshire north to the Orkneys, growing, but rarely in quantity, on rocky ledges and mountainsides from 300 m (1000 ft) to almost 900 m (3000 ft) in Caenlochan and on Meall Ghaordie. Outside Britain it has a similar distribution to that of *S. arbuscula* L., occurring widely in Scandinavia and eastwards to the Urals. Older records from the Alps and Pyrenees, and from elsewhere in central Europe are nowadays regarded as misidentifications of the allied *S. alpina* Scop. and *S. breviserrata* Flod.

The varieties listed by Boswell-Syme (*28* p. 256–257) are of little significance; var. *serrata* Syme and var. *procumbens* (Forbes) Syme seem to be indistinguishable from the typical plant; var. *arbutifolia* Syme is obscure, nor does the brief description furnish any satisfactory grounds for discrimination.

Male specimens of *S. myrsinites* are very rarely collected, probably because the catkins are mostly shed by the time botanists begin their investigations in the Highlands; there is no evidence that the male plant is any less common in nature than the female.

A: leaves × 2/3; B: ♂ catkins × 2/3; C: detail × 8; D: detail × 12; E: ♀ catkins × 2/3; F: detail × 8.

Salix myrsinites

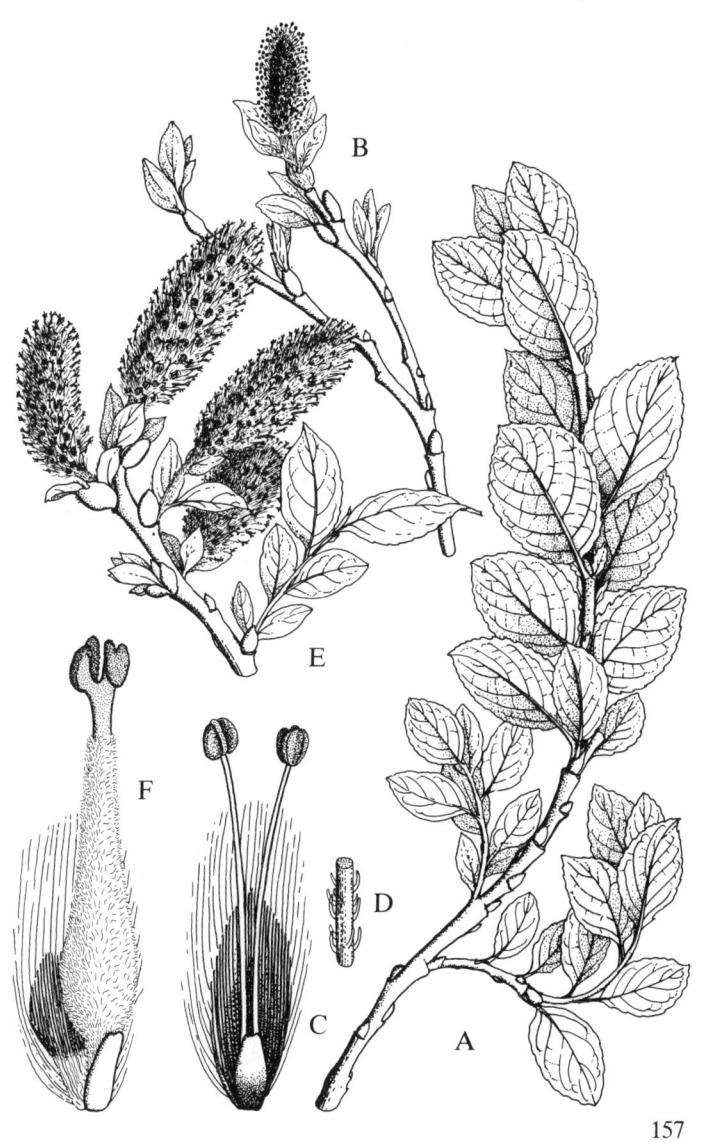

22. Salix herbacea L.

Dwarf Willow

The smallest of the British willows, generally less than 6 cm high, forming loose, flattened mats, with extensive, much-branched, rooting, underground stems, the terminal, exposed branches at first thinly pilose, soon becoming glabrous and dark shining brown or reddish; buds reddish, bluntly ovoid, seldom more than 3 mm long and 2 mm wide, at first thinly hairy or subglabrous, soon glabrous; leaves broadly and bluntly obovate or suborbicular, sometimes rather broader than long, very variable in size, ranging from 0.3–2 cm (or exceptionally 3 cm) long, and almost as wide or even a little wider, thinly white-pilose above and below at first, but soon becoming glabrous and dark shining green, venation prominently reticulate, margins rather evenly crenulate-serrate or rarely almost entire, apex rounded or shallowly emarginate, occasionally subacute, base rounded, subcordate or very broadly cuneate; petiole very short, usually less than 4 mm long, channelled above, white-pilose but becoming subglabrous; stipules minute, membranous and caducous, or wanting. Catkins appearing with the leaves in June and July (or as late as August at very high elevations) inconspicuous (usually less than 12-flowered), commonly terminating the short branches, subsessile or with hairy peduncles up to 8 mm long; bracts wanting; catkin-scales yellowish or more commonly stained reddish, oblong or obovate, 1–2 mm long, 0.5–0.8 mm wide, subglabrous or with a few hairs in the upper part, apex obtuse or sometimes shortly emarginate. Male catkins with 2 free stamens; filaments glabrous, 2–3 mm long; anthers broadly oblong, 0.8 mm long, 0.5 mm wide, yellow or tinged red; nectary variable, commonly cup-like and deeply lobed, surrounding the bases of the filaments, or sometimes consisting of 1 or 2 entire or lobed, distinct nectary-scales. Female catkins usually more conspicuous than male; ovaries narrowly flask-shaped, glabrous or rarely with a few hairs, sessile or very shortly stipitate, 2–3 mm long, 1 mm wide, often turning vinous-red with age; style short but distinct, stigmas short, spreading, deeply bifid; nectaries 1 or 2, oblong and entire or irregularly lobed. Capsule reddish-purple, up to 8 mm long and 3 mm wide.

A: habit × 2/3; B: larger leaves × 2/3; C: ♂ catkins × 2/3; D: detail × 12; E: ♀ catkins × 2/3; F: detail × 12.

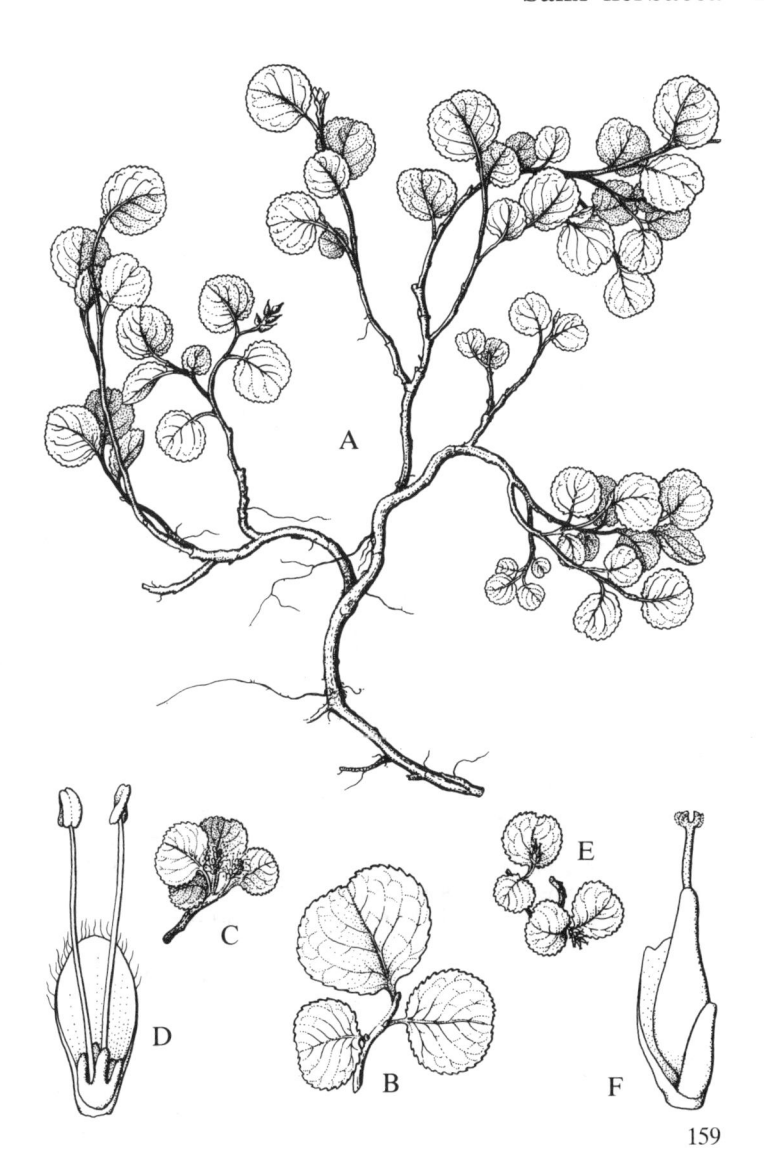

Salix herbacea 22

Salix herbacea is generally a plant of high altitudes in our area, reaching 1300 m (4300 ft) on Ben Nevis, and becoming rare below 600 m (2000 ft), though it is recorded from 150 m (500 ft) in the Shetlands. It has a wide distribution in Britain and Ireland, usually growing on moist, exposed rock ledges and rocky summits, or forming an alpine turf with grasses and sedges, often so prostrate and small that it is easily overlooked. In exceptionally sheltered sites, on moist screes, the stems can exceed 6 cm in length and the leaves 2 cm in width, but size apart, *S. herbacea* is remarkably uniform. It has a wide circumpolar distribution in Europe, Asia and N. America, and is also found on the Pyrenees and on the mountain ranges of central Europe.

22 × 17. **Salix herbacea** L. × **S. repens** L. = **Salix × cernua** E. F. Linton

A dwarf prostrate or ascending shrub, generally less than 10 cm high, with a tough, woody, much-branched, creeping rootstock; young shoots thinly or rather densely pubescent, soon glabrescent, twigs slender, glabrous or subglabrous, reddish; leaves ovate-oblong or obovate-oblong, 0.5–1(–2) cm long, 0.3–0.9(–1.2) cm wide, generally glabrous and sublustrous green above, paler or glaucescent below and thinly or rather densely clothed with adpressed hairs, apex shortly acute, obtuse, rounded or shallowly emarginate, base usually rounded, margins entire or minutely and regularly crenate-serrate; petiole very short, usually less than 3 mm long; stipules minute or wanting; catkins produced with the leaves in May and June, terminal on short, leafy, lateral shoots, shortly pedunculate, 15–20 mm long, 5–7 mm wide; catkin-scales oblong, obtuse or rounded at apex, yellowish or tinged red in the upper part, subglabrous with distinctly ciliate margins, about 1–2 mm long, 0.6–1 mm wide; male catkins with 2 free stamens, filaments glabrous, to 3 mm long; anthers broadly ovate-oblong, yellow or tinged reddish, about 0.7 mm long, and almost as wide, nectaries 2, the adaxial broadly oblong, about 0.5 mm long and about as wide,

---

A: habit and ♀ catkins × 2/3; B: ♂ catkins × 2/3; C: detail × 12; D,E: details of ♀ catkins × 12.

Salix × cernua 22 × 17

the abaxial slightly longer and narrower, sometimes bifurcate almost to base; female catkins with a shortly stipitate, narrowly flask-shaped, glabrous or adpressed-pubescent ovary 2–3 mm long, 1 mm wide, style short but distinct, stigmas 2, usually 2-cleft into 4 short narrow lobes. Capsule 3–4 mm long, apparently reddish, perhaps sterile.

Probably commoner in Scotland than would appear from the records, and noted from Mid Perth, S. Aberdeen, the island of Rhum, Ross, Argyll, Sutherland and Caithness, generally growing amongst *Calluna, Empetrum, Salix herbacea, Arctostaphylos*, etc. on rocky moorland. E. F. Linton thought *S. herbacea* × *S. repens* was a British endemic, but it has since been recorded from Norway (Floderus specimen, Kew) and might be expected to occur elsewhere in northern Europe. Male and female plants seem to be equally frequent in Britain. The Caithness specimens (Dunnet Hill, 1972, J. K. Butler, Kew) are more vigorous, pilose and larger-leaved than is usual in the hybrid, and show some approach to the mysterious *S.* × *grahamii* (p. 126).

23. Salix reticulata L.

Net-leaved Willow

A low shrub, forming loose open mats with extensive, much-branched, rooting underground stems, the exposed portions erect or ascending 5–15(–20) cm long, at first thinly hairy with long, silky hairs, but soon becoming glabrous and dark reddish-brown, buds rather large and conspicuous, oblong, blunt, 4–7 mm long, 3–4 mm wide, at first often rather densely villose, becoming glabrous and reddish-brown; leaves broadly and bluntly oval or suborbicular, 1.2–4(–5) cm long, 1–2.5(–3.5) cm wide, at first densely villose on both surfaces, with long subadpressed silky hairs, soon becoming glabrous and dull, dark green above, glabrous or subglabrous and uniformly pale ashy-grey below; nervation conspicuously reticulate, deeply impressed above and prominent below, margins entire,

---

A: habit × 2/3; B: ♂ catkin × 2/3; C: detail × 12; D: ♀ catkin × 2/3; E,F,G: details × 12.

Salix reticulata

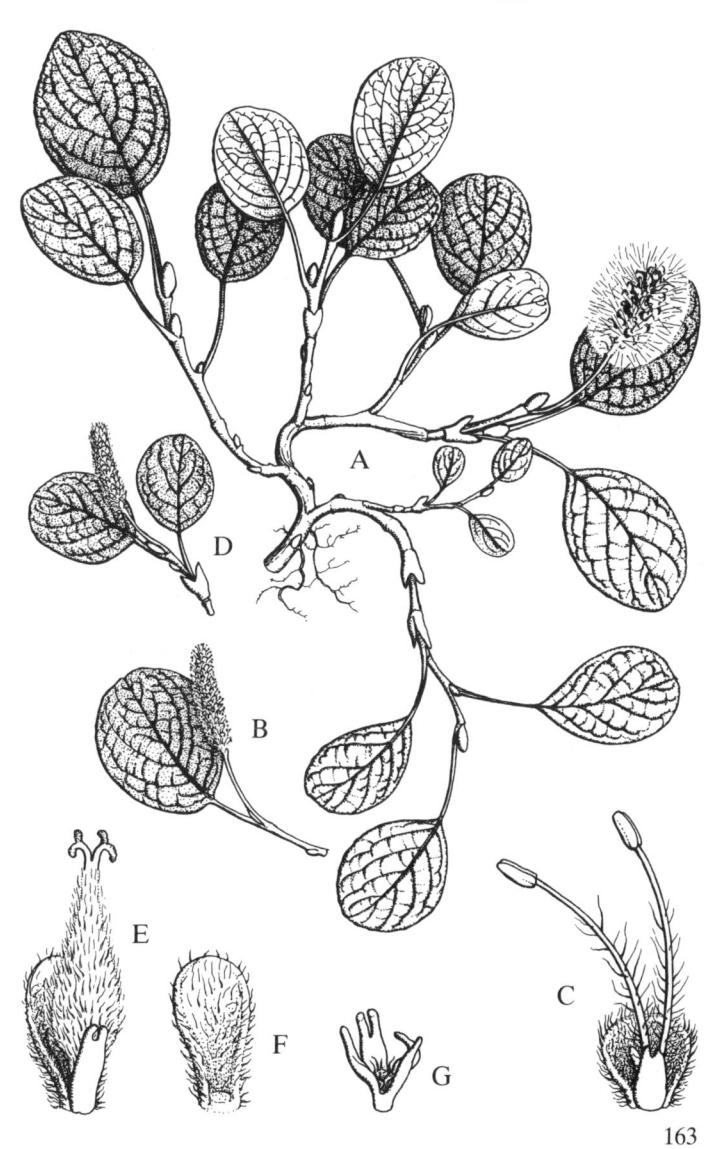

sharply recurved, minutely glandular, base rounded or broadly cuneate; petiole long and slender, usually exceeding 1 cm, sometimes 4 cm, or more, thinly villose at first, but soon glabrous, reddish and distinctly channelled above; stipules wanting, but leaf-bases often surrounded by the persistent, usually 2-lobed, bud-scale. Catkins narrowly cylindrical, 20–35 mm long, 4–5 mm wide, appearing with the fully developed leaves in late June or early July, erect, subterminal, leaf-opposed, with a long, arcuate, glabrescent peduncle up to 5 cm long; catkin-scales broadly obovate or suborbicular, about 1 mm long and nearly as wide, brownish or purplish, rather densely clothed, especially towards the base, with short, spreading, greyish hairs; stamens 2, free, filaments 1.5–3 mm long, distinctly pilose in the lower half; anthers broadly oblong or suborbicular, about 0.7 mm long, 0.6 mm wide, reddish or purplish before dehiscence; female catkins often rather more robust than the male; ovaries broadly flask-shaped, about 2 mm long, 1 mm wide, densely white-woolly; style short, but distinct; stigmas spreading, about 0.5 mm long, shortly bifid; nectaries of both male and female flowers variable, usually forming a cup around the base of the filaments or ovary, often irregularly laciniate. Capsule broadly ovoid, 5–6 mm long, 3–4 mm wide, hairy even when fully mature; old, dehisced fruiting catkins commonly persisting until June or July of the succeeding year.

Without question the most distinct and most easily identified of all the British Willows, as well as being one of the most attractive. *Salix reticulata* is, unfortunately, rare, being recorded only from Aberdeen, Angus, Sutherland and Inverness, and only to be seen in local abundance at altitudes above 2000 feet on Ben Lawers and adjacent mountains in Breadalbane, Perthshire, where it was first noted by Lightfoot in 1777. It grows on wet, often slightly calcareous, rocks and ledges, and has a circumpolar, boreal distribution across arctic and subarctic Europe, Asia and N. America. In Europe it extends south to the Pyrenees, Alps, Carpathians and mountains of Macedonia. Apart from a few slight variations in leaf shape and indumentum, it is uniform and unmistakable throughout its range.

24. Populus alba L.

White Poplar, Abele

A robust tree with a spreading crown, usually 15–20 m high in the British Isles, said to attain 40 m or more on the Continent; bark greyish, often fissured and blackish towards the base of the trunk, smooth above and commonly marked with horizontal, moniliform rows of conspicuous blackish lenticels; young shoots densely white-tomentose, the tomentum often persisting for a year or more, but the twigs ultimately glabrous, dark brown and sometimes rather glossy. Buds shortly ovoid, subacute, about 5 mm long, 3–4 mm wide, scales ovate, bluntish, tomentose at base, glabrescent and dark brown towards apex. Leaves of two kinds: those on short lateral spurs and at the base of long, leading shoots broadly and bluntly ovate, irregularly and bluntly sinuate-lobed, glabrescent or thinly tomentose; those towards the apex of the long shoots often deeply palmatilobed (like an *Acer*), the lamina 3–9 cm long, 3–10 cm wide, thinly floccose at first on the upper surface, becoming glabrous and dark green with age, persistently and strikingly white-tomentose below, lobes acute, obscurely and irregularly dentate, nervation not very prominent; petiole subterete, 5–6 cm long, tomentose. Catkins appearing in late February and March, well in advance of the leaves; catkin-scales pale brown, narrowly ovate or obovate-cuneate with an irregularly erose-dentate, long-ciliate apex. Male catkins usually more robust than female, 4–7 cm long, 0.8–1 cm wide; stamens 5–10; anthers purple. Female catkins 3–5 cm long, 0.6 cm wide at anthesis, lengthening considerably in fruit; ovary ovoid, glabrous, about 1.5 mm long, 1 mm wide; stigmas greenish, cleft almost to the base into 4 linear lobes 1–1.5 mm long. Capsule about 3 mm long, 2 mm wide, dull pale brown, shortly pedicellate, dehiscing by 2 valves.

Not infrequently planted, especially in the south, and sometimes forming dense colonies, arising from the sucker shoots which spread far from the parent tree, and make the White Poplar a nuisance to gardeners, although its two-coloured leaves are conspicuously decorative. The male tree is evidently extremely rare in Britain and

---

A: leaves × 2/3; B: ♂ catkins × 2/3; C: details × 8; D: ♀ catkins × 2/3; E: details × 8.

Populus alba 24

Ireland. Elwes and Henry (*8* p. 777) record that the only male flowers they could find were from a small and abnormal tree at Kew, and those illustrated here are taken from Swiss specimens.

Populus alba is widely distributed in central and S.E. Europe eastwards to Central Asia. It is, however, a fairly obvious introduction in many areas.

24 × 25. Populus alba L. × P. tremula L. = Populus × canescens (Ait.) Sm.

Grey Poplar

A tall, robust tree, said to attain 50 m, and commonly 20–30 m high; trunk greyish as in *P. alba,* often with horizontal, moniliform rows of blackish lenticels; young shoots at first densely white-tomentose, soon becoming glabrous and finally dark grey-brown; twigs usually stouter with more prominent bud-scars than in *P. alba.* Buds ovoid, subacute; scales tomentose at the base as in *P. alba.* Leaves of two kinds: those of the short spurs and at the base of the long, leading shoots suborbicular, like *P. tremula,* 3.5–6(–9) cm diam., glabrous or subglabrous on both surfaces, with sinuate-dentate margins and rather prominent reticulate venation; those of the upper parts of leading shoots often broadly ovate, 6–8(–10) cm long and about as wide, glabrescent and dark green above, thinly greyish-tomentose below, the tomentum thinning with age but seldom completely vanishing, margins coarsely and irregularly erose-dentate, sometimes shortly lobed; petiole of the short-shoot leaves often distinctly compressed laterally and glabrous, as in *P. tremula;* petioles of the leading-shoot leaves floccose-tomentose, subterete. Catkins appearing in March, well in advance of the leaves, the males common, the females occasionally found, but very rare. Male catkins resembling those of *P. tremula,* cylindrical-caudate, 4–9 cm long, 1–1.2 cm wide; catkin-scales brown, flabellate, 3–4 mm long and almost as wide, apex long-ciliate, coarsely and irregularly erose-dentate, but not as deeply incised as in *P. tremula;* disk oblique-entire; stamens 8–15, anthers reddish. Female catkins slender, 4–6 cm long at anthesis, about 0.5 cm wide; catkin-scales

---

A: leaves × 2/3; B: young leaves × 2/3; C: stipule × 3; D: ♂ catkins × 2/3; E: details × 8; F: ♀ catkins × 2/3; G: details × 8.

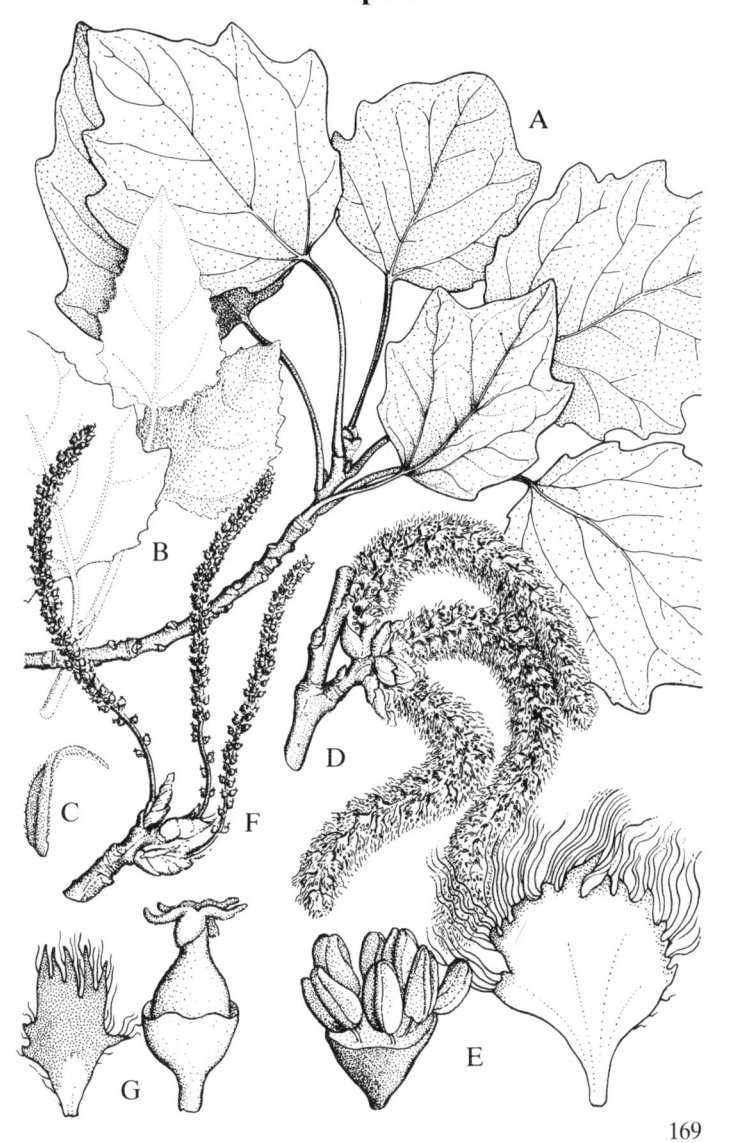

Populus × canescens 24 × 25

rather small and narrow, irregularly laciniate, thinly ciliate; disk glabrous; stigmas yellowish or pinkish, deeply 2-cleft or divided into 4 (or more) linear lobes about 0.8 mm long. Ripe capsules not seen.

After protracted controversy, it is now generally agreed that *Populus* × *canescens* is a hybrid between the White Poplar and the Aspen, a conclusion supported by the fact that it shares the characters of the presumed parents. Hybrids produced by artificial crosses agree closely with the plant found in our area, and it has been shown that the hybrid is capable of producing viable seeds, and can be back-crossed with either parent (see S. Bartkowiak and S. Białobok in *Arboretum Kórn.*, 11: 105–151 (1966)). It is just possible that some of our specimens may have arisen from hybridization in this country, but in most cases *P.* × *canescens* is a fairly obvious introduction, probably propagated clonally from suckers (*P.* × *canescens*, like *P. alba*, is difficult to root from ordinary cuttings), with the male plant predominating, though females have been collected in S. Devon, N. Somerset, W. Gloucestershire, Surrey, E. Kent, Hertfordshire and Bedfordshire. Elwes and Henry (*8* p. 1779 footnote) may be correct in suggesting that the 10,000 Abeles imported into England from Holland some time before 1641, were in fact *P.* × *canescens*. The hybrid is common in the south-eastern half of England, rather infrequent elsewhere. Abroad it has a more northern distribution than *P. alba*, but occurs from France, Belgium and Holland east to the Caucasus.

25. **Populus tremula** L.

Aspen

A suckering tree up to 20 m high, usually with a broad, much-branched crown, but sometimes no more than a shrub in exposed or impoverished sites; bark smooth, greyish, often dark and fissured towards the base of the trunk; shoots glabrous or thinly hairy, soon becoming glabrous; twigs dull greyish-brown, often rather gnarled with prominent protruding leaf scars; buds ovate,

---

A: leaves × 2/3; B: sucker leaves × 2/3; C: ♂ catkins × 2/3; D: details × 6; E: ♀ catkins × 2/3; F: details × 6.

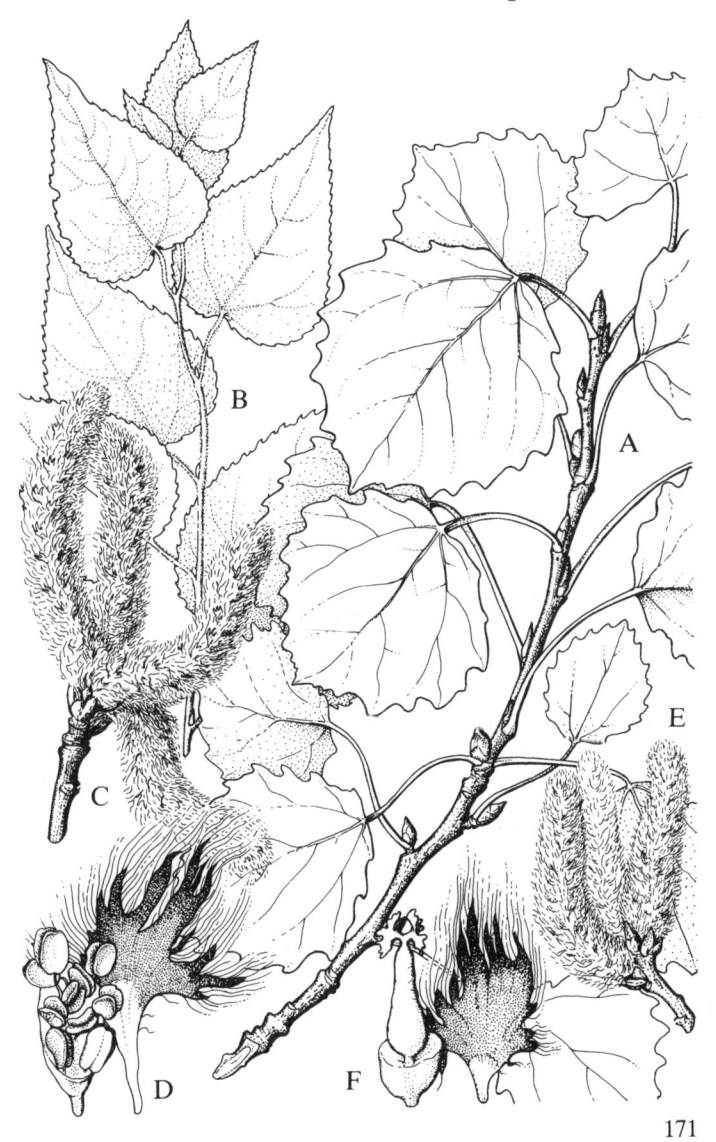

Populus tremula

acute, those of sterile growths about 5 mm long, 3 mm wide, those containing catkins often large and conspicuous, about 8–10 mm long, 6–7 mm wide, scales shining rich brown, becoming sticky as the buds expand in the spring. Leaves of sucker shoots ovate-cordate, 5–12 cm long, 4–9 cm wide, thinly pilose above and below, margins irregularly serrate, apex acute, petiole subterete, pilose, 1.5–3 cm long; leaves of normal growths sometimes adpressed-sericeous at first, but soon glabrous, broadly ovate, suborbicular or oblate-orbicular, 1.5–8 cm long, and about as wide or a little wider, dark green above, slightly paler below, base truncate or shallowly cordate, margins coarsely, bluntly and rather irregularly sinuate-serrate; petioles usually long and slender, commonly 4–7 cm long, glabrous, strongly compressed laterally. Catkins appearing in late February and March, well in advance of the leaves, cylindrical-caudate, 5–8 cm long, 1–1.5 cm wide, often crowded towards the tips of the year-old twigs; catkin-scales flabellate, 5–6 mm long, 4–5 mm wide, dark brown, densely clothed externally with long white hairs, glabrous internally, apex deeply and irregularly laciniate. Male flowers with an oblique entire disk and 6–12(–15) stamens with short filaments, and rather large, oblong, crimson-purple anthers. Female flowers with a narrowly flask-shaped, glabrous, scabridulous ovary about 3–4 mm long, 1–1.5 mm wide; stigmas purplish or pinkish, divided into 2 or more rather irregular lobes less than 0.5 mm long. Capsules small, not much exceeding 4 mm in length and 1.5 mm in width; shortly pedicellate, seated in a shallow, cupuliform disk.

The most widely distributed of Poplars, ranging across temperate Europe and Asia to China and Japan, and from the Mediterranean north to the limit of tree vegetation in the Arctic tundra. In the southern part of its range *P. tremula* is restricted to montane woodland; further north it descends to the lowlands and is a common component of scrub on poor, sandy or peaty soils; isolated specimens frequently occur as chasmophytes on the steep rocky sides of mountain streams. It is on record from almost every vice-county in Britain and Ireland.

Surprisingly, in view of its immense distribution, the Aspen does not vary much: *P. tremula* L. var. *villosa* (Lang) Wesm. has the shoots and leaves at first densely, and sometimes persistently, clothed with silky hairs; the extreme can look distinct, but is connected to typical, glabrous *P. tremula* by numerous intermediates. *P. tremula* L. f. *microphylla* A. Brown ex Schneid. has uniformly small leaves, scarcely more than 2.5 cm wide. It can look very distinct, but may be merely a juvenile form, or habitat state. The pendulous *P. tremula* L. var. *pendula* Loudon seems to survive only as a horticultural curiosity.

26. Populus nigra L.

Black Poplar

A robust tree sometimes 30–35 m high, with a broad, rounded crown and massive, often downcurved branches; trunk covered with coarsely fissured dark, greyish bark, often disfigured with large swellings or burrs; twigs glabrous or glabrescent, terete and rather lustrous ochre-brown; buds viscid, dark shining brown, narrowly ovate-acuminate, the terminal sometimes 1 cm long, 0.5 cm wide, the laterals rather shorter and sometimes distinctly curved outwards from the twig, bud scales acute, often rather loosely imbricate. Leaves deltoid-ovate, 5–10 cm long, 3–9 cm wide (sometimes larger on robust sucker or coppice shoots), dark, sublustrous green above, slightly paler below, apex elongate-acute or shortly acuminate-caudate, base broadly cuneate or almost truncate, rarely rounded, margins narrowly translucent, distinctly or indistinctly, and usually bluntly serrate, with the acumen frequently entire or subentire; nervation rather prominent below, with widely spreading, upcurved lateral nerves; petiole laterally compressed, 3–7 cm long, without glandular swellings at the point of junction with the lamina. Catkins appearing before the leaves in late March or April, caudate-cylindrical, rather lax, 3–5 cm long, 0.6–0.7 cm wide at anthesis, shortly but usually distinctly pedunculate, the females elongating to 9–10 cm in fruit; rhachis glabrous or very sparsely pubescent; catkin-scales soon deciduous, membranous, greenish or brownish, 1–2 mm long and about as wide, apex deeply laciniate into narrow lobes. Male flowers with an oblique, shallow, cup-shaped disk about 1.5 mm diam.; stamens 12–15(–20); filaments very slender, about 0.7 mm long; anthers crimson, oblong, about 1 mm long, 0.8 mm wide. Female flowers with a subglobose, glabrous ovary about 1.5 mm diam. seated in a glabrous cupuliform disk about 2 mm diam.; stigmas 2, pale greenish, deeply bifid apically and sometimes basally into narrow, blunt lobes. Capsules 5–6 mm long, 4 mm wide, broadly ovoid, dehiscing by 2 recurved valves.

A recent survey, conducted by Mr E. W. B. H. Milne-Redhead, has shown that the Black Poplar is widely, but often sparsely distributed along river valleys in most of southern England, south of a line joining the Mersey and the Humber. Within this area it is absent (except possibly where planted) only in Cornwall and extreme S.W. England, and in western Wales. North of the line it may occur as a planted tree but cannot be considered indigenous. The British trees, as well as many from France, western and southern Germany, belong to *P. nigra* L. var. *betulifolia* (Pursh) Torrey (*P. betulifolia* Pursh; *P. nigra* L. ssp. *betulifolia* (Pursh) W. Wettst.), which differs from the type only in the thin (and deciduous) pubescence of the young shoots, petioles and rhachises. The variety was first recognised as distinct by N. American botanists, and was, indeed, thought to be an American tree, though it had in fact been originally imported from Europe.

Outside Britain, *Populus nigra* (in the aggregate sense) is widely distributed in central and southern Europe, the Mediterranean region and eastwards to Central Asia (Kazakhstan). The familiar, fastigiate Lombardy Poplar (*P. nigra* L. var. *italica* Muenchh.) is now (Bugała, 1967) considered to have orginated in northern Italy at the end of the 17th or in the early 18th century, and to have reached Britain about 1758. Typically it is a male, propagated clonally; occasional records for female plants most probably involve hybrids between *P. nigra* var. *italica* and some other species, or are misidentifications of *P. nigra* L. var. *afghanica* Aitch. et Hemsl., a distinct, but very similar fastigiate variant commonly grown in the eastern Mediterranean region and N. Africa. Bugała refers this latter to a distinct species, as *P. uzbekistanica* Komarov cv. *Afghanica*.

A: leaves × 2/3; B: ♂ catkins × 2/3; C: details × 6; D: ♀ catkins × 2/3; E: details × 6.

Populus nigra 26

27a. Populus deltoides Marsh. × P. nigra L. = **Populus × canadensis** Moench var. **serotina** (Hartig) Rehder

A very large tree, often more than 30 m high, usually with a long, unbranched trunk, and a wide, fan-shaped crown; trunk covered with coarsely but rather regularly fissured grey-brown bark, without the burrs found in *P. nigra*; branches erect or erecto-patent, not downcurved; twigs glabrous, shining olive-grey, often somewhat angled; buds shining greenish-brown, viscid, balsamic, narrowly ovoid-acuminate, 1–2 cm long, 0.4–0.7 cm wide, often distinctly curved outwards from the branch, scales rather few and large, acute, loosely imbricate. Leaves deltoid-ovate, 6–10 cm long, 4–10 cm wide, bright lustrous green above, slightly paler below, glabrous, apex elongate-acute, base almost truncate or shallowly cordate, sometimes rounded, margins strongly but bluntly serrate, narrowly translucent; nervation rather prominent below, with widely spreading, upcurved lateral nerves; petioles glabrous, strongly compressed laterally, 4–10 cm long, often bearing one or two small glands at the point of junction with the lamina, but sometimes eglandular. Catkins appearing well in advance of the leaves in late March or early April, narrowly ovoid-cylindrical, dense, 3–6 cm long, 0.8–1 cm wide, shortly pedunculate; rhachis glabrous; catkin-scales broadly flabellate, membranous, purplish-tipped, 4–5 mm long and as wide or a little wider, apex deeply fimbriate-laciniate; disk shallow, cup-shaped, glabrous, about 3.5 mm diam.; stamens crowded, 20–25 or more; filaments about 0.8 mm long; anthers narrowly oblong, about 1.5 mm long, 0.8 mm wide, crimson.

One of the commonest planted Poplars, and the best known of the "Euro-American" hybrids, usually easy to recognise by its tall (often slanting) trunk and fan-shaped crown, dense red catkins and handsome reddish-bronze young foliage. The truncate or subcordate leaf-bases, usually (at least in some of the leaves) bearing one or two small glands at the junction with the petiole, will also assist in distinguishing it from *P. nigra*. Only the male tree is known; it is generally supposed to have arisen in France some time about the middle of the 18th century and to have reached our islands a little

---

A: leaves × 2/3; B: catkins × 2/3; C: details × 6.

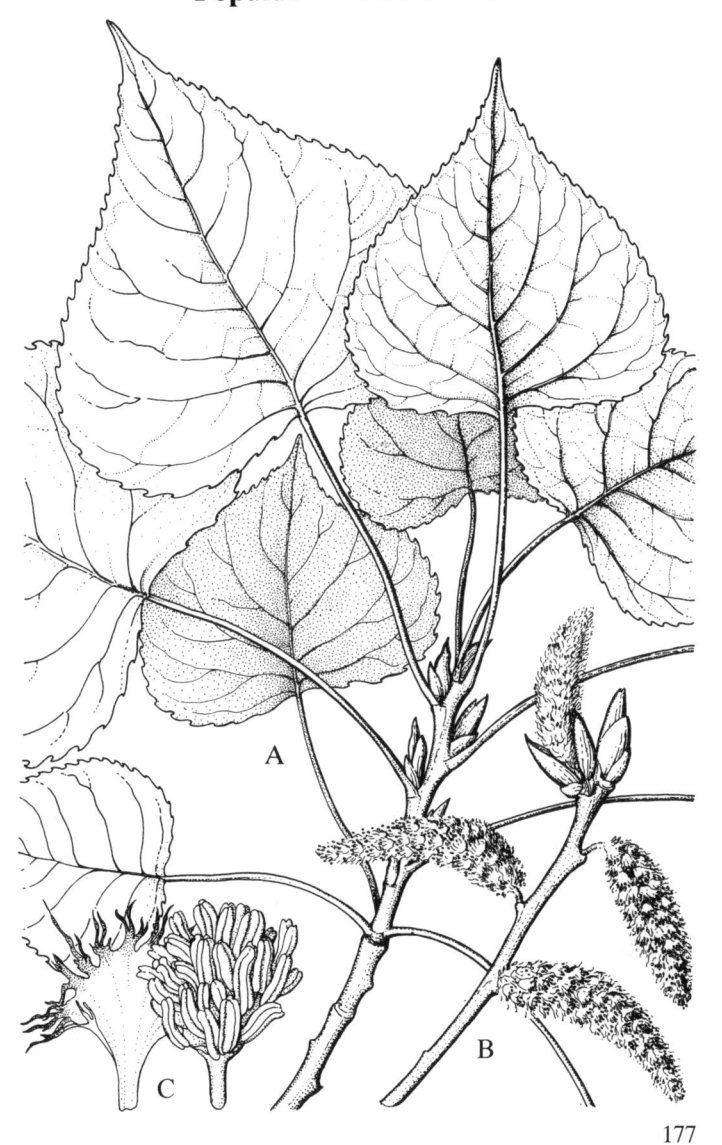

Populus × canadensis var. **serotina** **27a**

later in the same century. It is now probably to be seen in every county, and, in some areas, has quite replaced *P. nigra*. The "Euro-American" group consists of hybrids between the European *P. nigra* L. (p. 173) and *P. deltoides* Marshall from eastern North America. The latter, brought to Europe in the early 19th century, is now very rare in cultivation, having been almost wholly replaced by hybrids with *P. nigra*, of which there are said to be at least thirty-five named or distinct nothomorphs (and now probably many more) in cultivation (Bugała, 1973). True *P. deltoides* (*P. monilifera* Ait.) resembles *P. × canadensis* var. *serotina* but is a more massive, stocky tree with thick, spreading branches, very coarsely serrate leaves with ciliate margins, and with glands regularly present at the apex of the petiole.

27b. Populus nigra L. × P. × canadensis Moench var. serotina (Hartig) Rehder = Populus × canadensis Moench var. marilandica (Bosc ex Poir.) Rehder

A large tree, up to 30 m high, trunk seldom as long or as straight as in var. *serotina*, covered with coarsely fissured bark, but without burrs; crown broadly domed, the branches numerous, spreading, not erect or downcurved; twigs yellowish-grey, terete, glabrous, rather glossy; buds shining pale brown, viscid, balsamic, narrowly ovoid-acuminate, 1–1.4 cm long, 0.3–0.5 cm wide, distinctly smaller than those of *P. × canadensis* var. *serotina*, scales rather few, acute, not very closely imbricate. Leaves deltoid- or rhomboid-ovate, 4–12 cm long, 4–9 cm wide, pale yellowish-green when unfolding and often sparsely pubescent especially near the translucent margin, glabrous and sublustrous bright green when mature, apex cuspidate-acuminate, base broadly cuneate or almost truncate, margins rather coarsely serrate with acute or subacute, forward-pointing teeth; petioles glabrous, laterally compressed, 4–10 cm long, one or two glands sometimes (but by no means always) present at the junction of petiole and lamina. Catkins appearing before the leaves in late March and early April, shortly pedunculate, 5–6 cm long at anthesis, elongating to 12 cm or more and

---

A: leaves and fruits × 2/3; B: older ♀ catkins × 2/3; C: younger ♀ catkins × 2/3; D: details × 6.

Populus × canadensis var. marilandica 27b

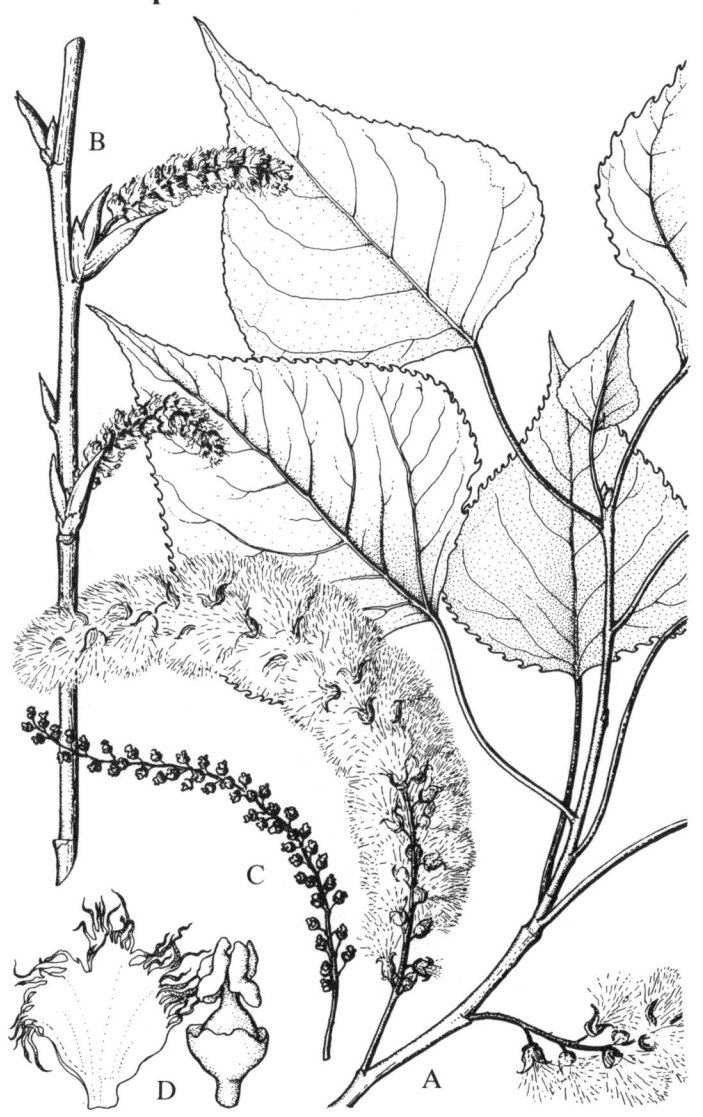

becoming very lax in fruit; rhachis glabrous; catkin-scales broadly flabellate, membranous, glabrous, pale brownish or greenish with closely fimbriate-laciniate margins; flowers with spreading pedicels about 1 mm long; disk shallowly cup-shaped, about 2 mm diam.; ovary glabrous, subglobose, about 1.2 mm diam.; style short, stout; stigmas 2–4, yellowish green, with deflexed, deeply bifid lobes. Capsule about 5 mm long, 3 mm wide, dehiscing by 2–4 valves and shedding an abundance of cottony seeds. Male plants unknown.

Said to have arisen some time around 1800, probably in Holland, and to have spread rapidly to other parts of Europe. It is less frequently planted in Britain and Ireland than *P.* × *canadensis* var. *serotina*, partly because it is inferior as a timber tree, and partly because the copious floss shed by its fruits can be a serious nuisance, especially in built-up areas.

P. × *canadensis* Moench var. *eugenei* (Simon-Louis ex Mathieu) Rehder, always male, resembles var. *marilandica* in leaf, but has a narrow crown, with short ascending lateral branches; it may have originated as a hybrid between one of the "Euro-American" Poplars (*P.* × *canadensis* Moench var. *regenerata* (Henry) Rehder) and the Lombardy Poplar (*P. nigra* L. var. *italica* Muenchh.).

Other hybrid Poplars of the same or similar ancestry are: *P.* × *canadensis* Moench var. *regenerata* (Henry) Rehder, believed to be a hybrid between *P.* × *canadensis* var. *marilandica* and *P.* × *canadensis* var. *serotina,* and resembling the latter parent, but female, with a broader, more open crown, and less erect branching; the unfolding leaves are in shades of green and not reddish-brown as in *P.* × *canadensis* var. *serotina*. It is said to have arisen at Arcueil in France, in 1814, and was at one time so frequently planted to screen railway goods-yards as to earn the name "Railway Poplar". Although female, it is largely sterile, and, unlike var. *marilandica,* sheds only a small quantity of floss.

P. × *robusta* Schneid., reckoned a hybrid between the N. American *P. angulata* Ait. f. *cordata* (David ex Thomas) Rehder (often considered a variety of *P. deltoides* Marsh.) and *P. nigra* L. var. *plantierensis* Schneid., is known to have arisen in 1895 in the nursery of Simon-Louis at Plantières near Metz (N.E. France). It is a male tree, with minutely pubescent shoots and dark reddish-brown unfolding leaves. *P.* × *robusta* is very vigorous, with a clean, straight trunk, a narrow crown, and short ascending lateral branches, and has in recent years become popular with commercial growers.

28. Populus candicans Ait.
(Populus × gileadensis Rouleau)

Balsam Poplar

A medium-sized, suckering tree, usually attaining 20–25 m at maturity in this country, but often much smaller; crown rather broad, with spreading and suberect branches; bark smooth and greenish-grey on juvenile specimens, becoming coarsely and broadly fissured in mature trees; shoots at first angular and thinly pubescent, becoming subterete and glossy dark brown with age; buds narrowly ovoid-acuminate, 10–13 mm long, 3–4 mm wide, glossy, viscid, strongly balsamic with a few acute, shortly pubescent scales. Leaves broadly cordate, 5–15 cm long (or more on sucker shoots), 4.5–12 cm wide, dark sublustrous green above, conspicuously whitish below with rather prominent nervation, apex rather abruptly acute-cuspidate, margins regularly but rather bluntly serrate with upcurved, finely ciliate teeth; petioles 3–7 cm long, pubescent, slightly flattened on the upper surface, but not laterally compressed, with 2 glands at or near the point of junction with the lamina. Female catkins (the male tree is unknown) appearing in late March shortly before the development of the leaves, caudate-cylindrical, very shortly pedunculate, at first 4–6 cm long, lengthening to 16 cm or more in fruit, peduncle and rhachis thinly pubescent; catkin-scales broadly flabellate, deeply pectinate-laciniate, pallid, glabrous, soon deciduous, about 3–4 mm long, and about as wide; flowers shortly and stoutly pedicellate, disk glabrous, shallowly cupuliform, about 2.5 mm diam. at anthesis, becoming larger with age; ovary small, glabrous, subglobose, at first almost concealed within the disk; style very short; stigmas 2, very broad, deflexed, irregularly lobulate. Ripe fruit and seeds not seen.

Like the following (*P. trichocarpa* Hooker) readily recognised by its strong, balsamic smell, which, especially on showery days in spring, when the buds are unfolding, will scent the air for yards around. *Populus candicans* must have been a commonly planted tree at one time, but, probably because of its prolific suckering, and perhaps because of its proneness to canker, it is not often seen nowadays in gardens, though it persists by moist roadsides or stream banks in almost every part of Britain and Ireland, often overlooked by the writers of local Floras, and sometimes confused with *P. trichocarpa*. It is supposed to have been introduced from

North America by Dr John Hope, some time around 1772, but is evidently not truly native in Canada or the U.S.A., though well known there (usually as Balm of Gilead) since the early colonial period. It has been reckoned a variety of *P. balsamifera* L. (*P. tacamahacca* Mill.), but this has ovate, subglabrous leaves, seldom markedly cordate at the base as in *P. candicans*. Others regard it as a hybrid between *P. balsamifera* L. and *P. deltoides* Marsh., but there is, as yet, no general agreement regarding its origin. Only the female clone is known in our area and, apparently, in America.

29. Populus trichocarpa Torrey & Gray ex Hooker

A tall, graceful tree, said to attain more than 60 m in western North America, but rarely exceeding 35 m in Britain; crown rather narrow and conical, with suberect or ascending branches; bark grey-brown, at first lightly fissured, but becoming rough in mature specimens; shoots at first sharply angular and shortly puberulous, soon becoming glabrous; older twigs terete or subterete, ochre-grey; buds narrowly ovoid-acuminate, 10–15 mm long, 3–4 mm wide, glossy, viscid, strongly balsamic, with a few, rather loose scales. Leaves broadly or narrowly ovate, 5–15(–23) cm long, 4–10(–16) cm wide, dark shining green above, conspicuously whitish below with prominent nervation, apex rather shortly acute, base broadly cuneate, truncate or very slightly cordate, margins shallowly and bluntly serrate with glandular teeth; petiole rather short and stout, seldom exceeding 4 cm, subterete or narrowly channelled along its adaxial surface, puberulous or glabrous, with or without 2 inconspicuous glands at or near the point of junction with the lamina. Female catkins appearing in late March and early April, a little before the foliage, caudate cylindrical, stout, very shortly pedunculate, 5–9 cm long, about 1 cm wide; peduncle and rhachis thinly hairy; catkin-scales broadly flabellate, deeply and irregularly laciniate, pale brown, thinly hairy externally, soon deciduous, about 3–5 mm long and as wide or wider; flowers shortly and stoutly pedicellate, disk shallowly cupuliform, glabrous, or shortly ciliate; stamens very numerous (30–60), filaments slender, filiform, pallid,

---

A: leaves × 2/3; B: young ♀ catkins × 2/3; C: details × 6; D: mature ♀ catkin × 2/3.

Populus candicans 28

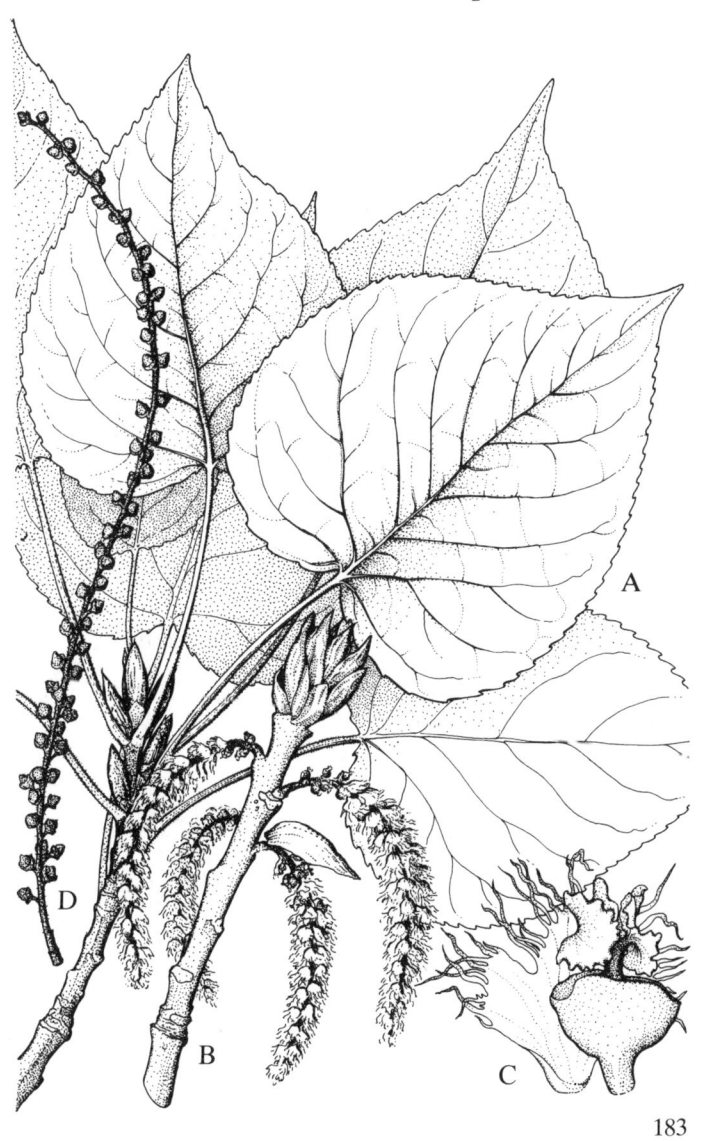

anthers oblong, crimson. Female catkins (rarely, if ever, seen in our area) with well-developed, hairy peduncles, 6–10 cm long at anthesis, elongating to 15–20 cm in fruit, catkin-scales as in male inflorescences; disk sessile or very shortly pedicellate, ovary globose, about 4 mm diam., densely grey-tomentose, stigmas 2, very broad, deflexed, irregularly lobulate.

Probably the most popular of the Balsam Poplars, and certainly the most handsome, growing with great rapidity to 15 m or more, often with a clean trunk and symmetrical branching. The buds and young foliage will fill a garden with a sweet balsamic fragrance, especially after spring showers. Although *Populus trichocarpa* is so commonly planted, and so readily propagated from suckers, it can scarcely be called a naturalized tree, and is certainly much less in evidence than *P. candicans* Aiton. It is said to be very susceptible to bacterial canker, and suckers less freely than many other *Populus* species and hybrids. Elwes and Henry (*8* p. 1837) say that the oldest specimen in our area is one planted in Edinburgh Botanic Garden in 1892. Only the male tree has been seen, though the female may have been introduced. All the British and Irish material differs from typical (Californian) *P. trichocarpa* in having longer, proportionately narrower leaves, with more noticeable marginal teeth and less cordate leaf-bases. It would appear to agree with var. *hastata* (Dode) Henry, from the northern part of the range of the species, but the validity of the distinction requires additional proof, and examination of a wider range of American specimens; *P. trichocarpa* is widely distributed along the Pacific seaboard, from California north to Alaska.

A: leaves × 2/3; B: ♂ catkins × 2/3; C: details × 6; D: ♀ catkins × 2/3; E: details × 6.

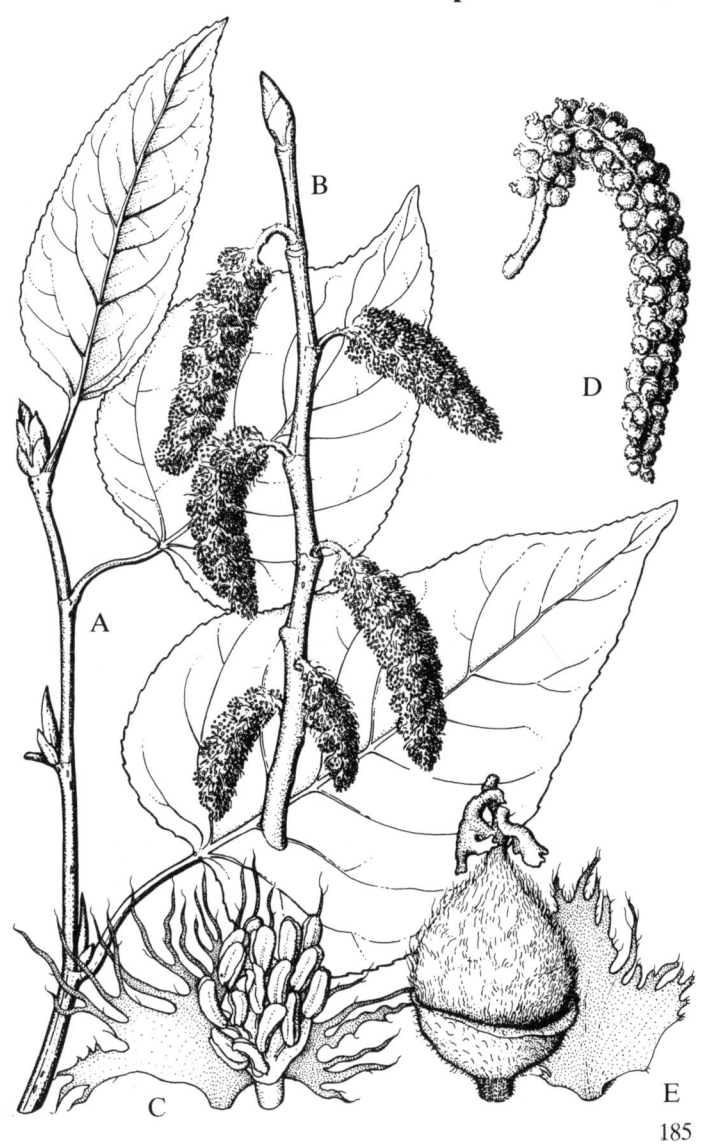

Populus trichocarpa

REFERENCES AND SELECTED BIBLIOGRAPHY

1. ANDERSSON, N. J. 1867. *Monographia Salicum*. Pars 1. Stockholm.
2. ANDERSSON, N. J. 1868. *Salicaceae* in De Candolle, A. P. *Prodromus Systematis Naturalis Regni Vegetabilis*, **16(2)**: 190–323. Paris, Strasbourg, London.
3. CAMUS, A. & CAMUS, E.-G. 1904–1905. *Classification des Saules d'Europe et Monographie des Saules de France*. Paris.
4. CHMELAR, J. 1976. *Die Weiden Europas*. Wittenberg Lutherstadt.
5. CLARKE, D. 1976. *Populus* in Bean, W. J. *Trees & Shrubs Hardy in the British Isles*, 8th ed., **3**: 293–328. London.
6. CLARKE, D. 1980. *Salix* in Bean, W. J. *Trees & Shrubs Hardy in the British Isles*, 8th ed., **4**: 246–312. London.
7. DRUCE, G. C. 1932. *The Comital Flora of the British Isles*, 271–276. Arbroath.
8. ELWES, H. J. & HENRY, A. 1913. *The Trees of Great Britain and Ireland*, **7**: 1743–1846. Edinburgh.
9. FORBES, J. 1829. *Salictum Woburnense*. London.
10. FRANCO, J. AMARAL. 1964. *Populus* in Tutin, T. G. et al. eds. *Flora Europaea*, **1**: 54–55. Cambridge.
11. FRASER, J. 1932. Some planted or cultivated willows. *Rep. B.E.C.* **9**: 719–721. Arbroath.
12. FRASER, J. 1933. Revised nomenclature of *Salix*. *Rep. B.E.C.*, **10**: 367–371. Arbroath.
13. GILBERT-CARTER, H. 1930. *Our Catkin-bearing Plants*, 4–20. Oxford.
14. GILBERT-CARTER, H. 1936. *British Trees and Shrubs*, 47–56. Oxford.
15. HÅKANSSON, A. 1955. Chromosome numbers and meiosis in certain Salices. *Hereditas*, **41**: 454–483. Lund.
16. LINTON, E. F. 1913. A monograph of the British Willows. *J. Bot., Lond.*, **51** suppl. London.
17. LOUDON, J. C. 1835–1838. *Arboretum et Fruticetum Britannicum*, **3**: 1453–1670; **7**: tt. 62 E–63 O. London.
18. MOSS, C. E. 1914. *Salicaceae* in *The Cambridge British Flora*, **2**: 13–68. Cambridge.

19 PERRING, F. H. & WALTERS, S. M. eds. 1962. *Atlas of the British Flora*. London.
20 RECHINGER, K. H. 1950. Observations on some Scottish willows. *Watsonia*, **1**: 271–275. Arbroath.
21 RECHINGER, K. H. 1957. *Salix* in Hegi, G. *Illustrierte Flora von Mittel-Europa*, 2nd ed., **3**: 44–185. Munich.
22 RECHINGER, K. H. 1964. *Salix* in Tutin, T. G. et al. eds. *Flora Europaea*, **1**: 43–54. Cambridge.
23 REHDER, A. 1940. *Manual of Cultivated Trees and Shrubs*, 2nd ed., 71–111. New York.
24 REHDER, A. 1949. *Bibliography of Cultivated Trees and Shrubs*, 65–87. Jamaica Plain, Mass.
25 SEEMEN, O. von, 1908–1910. *Salicaceae* in Ascherson, P. & Graebner, P. *Synopsis der mitteleuropäischen Flora*, **4**: 54–350. Leipzig.
26 SMITH, J. E. 1804. *Flora Britannica*, **3**: 1039–1072. London.
27 SMITH, J. E. 1828. *The English Flora*, **4**: 163–233. London.
28 SYME, J. T. B. ed. 1868. *English Botany*, 3rd ed., **8**: 200–261, tt. 1303–1379. London.
29 WADE, W. 1811. *Salices*. Dublin.
30 WARREN-WREN, S. C. 1972. *Willows*. Newton Abbot.
31 WHITE, F. B. 1890. A revision of British willows. *J. Linn. Soc. Bot.*, **27**: 333–457.
32 WIMMER, F. 1866. *Salices Europaeae*. Bratislava.

GLOSSARY

Abaxial: facing away from a central axis.
Accrescent: increasing in size.
Acumen: tip of an organ.
Acuminate: tapering to a point.
Acute: shortly pointed.
Adaxial: facing towards a central axis.
Adpressed: pressed against the surface.
Alternate: arising at different levels on a central axis.
Androgynous: with mixed male and female flowers.
Anthesis: flowering period.
Ascending: curving upward.
Attenuate: drawn out into a slender acumen.
Balsamic: with a balsam-like fragrance.
Bifid: 2-cleft.
Bract: reduced, modified leaf at base of catkin or along catkin-bearing shoot.
Bracteole: a secondary bract.
Caducous: falling off early.
Canaliculate: channelled.
Catkin-scale: small membranous bract subtending flower in *Salix*, *Populus* and other catkin-bearing plants.
Caudate: with a long, tail-like acumen.
Ciliate: fringed with hairs, like eyelashes.
Chasmophyte: growing in rock-crevices.
Clinal: with the characteristics of a cline.
Cline: a gradation of characters, usually over a geographic area.
Clonal: belonging to a clone.
Clone: an individual propagated vegetatively or asexually.
Connate: joined together.
Cordate: shaped like the conventional heart.
Coriaceous: leathery.
Crenate: with blunt, rounded teeth; scalloped.
Crenulate: minutely crenate.
Cuneate: wedge-shaped.
Cupuliform: like an acorn-cup.
Cuspidate: abruptly constricted at apex with a short, narrow acumen or *cusp*.

Cyathiform: cup-shaped.
Deciduous: falling off.
Decumbent: lying along the ground but with an upturned apex.
Dehiscence: bursting of a ripe capsule or anther.
Deltoid: more or less triangular.
Dentate: toothed.
Denticulate: with small teeth.
Dioecious: with male and female flowers on separate plants.
Discolorous: of 2 distinct colours.
Dorsally: on the back, or facing away from a central axis.
Ebracteate: without bracts.
Emarginate: indented at apex.
Entire: margin smooth, without teeth or lobes.
Erose: margin ragged as if gnawed.
Exserted: protruding.
Exstipulate: without stipules.
Falcate: sickle-shaped.
Flabellate: fan-shaped.
Floccose: covered with tufts of hair that rub off easily.
Foliaceous: resembling a leaf.
Free: not joined together.
Fuscous: blackish-brown.
Glabrescent: becoming glabrous.
Glabrous: without hair.
Glaucescent: somewhat glaucous.
Glaucous: bluish-grey.
Holt: ground set aside for the cultivation of osiers.
Indigenous: native.
Indumentum: hair-covering.
Laciniate: irregularly and jaggedly cut.
Lamina: leaf-blade.
Lanceolate: lance-shaped, broadest below the middle, and at least 3 times as long as wide.
Lanuginose: woolly.
Linear: long and narrow, with margins parallel or almost parallel.
Lobulate: with small or secondary lobes.
Lunate: shaped like a crescent moon.
Moniliform: constricted at intervals, like a string of beads.
Mucronate: terminating abruptly in a short point or *mucro*.
Mutant: a sport, or abrupt hereditary variation.
Nectary-scale: nectar-secreting gland or glands at the base of the stamens and pistils in *Salix*.
Nervation: system of primary nerves.

Oblong: of leaves, roughly rectangular, with the broadest part about the middle, and the margins sub-parallel or slightly convex.
Obovate: inversely ovate, with the broadest part above the middle.
Opposite: arising at the same level on opposing sides of a common axis.
Orbicular: round.
Ovate: more or less egg-shaped in outline, with the broadest part below the middle.
Ovoid: egg-shaped.
Palmatilobed: lobed from a common centre, like the fingers of a hand; digitate.
Patent: spreading.
Pectinate: closely pinnatifid, like the teeth of a comb.
Pedicellate: with a pedicel, or flower-stalk.
Pedunculate: with a peduncle, or inflorescence-stalk.
Pendulous: hanging down.
Persistent: not falling off readily.
Petiolate: with a petiole or leaf-stalk.
Pilose: hairy.
Polymorphic: represented by many forms or variants.
Prostrate: lying flat along the ground.
Pruinose: hoary; covered with a whitish "bloom".
Puberulous: minutely pubescent.
Pubescence: downiness.
Pubescent: downy, clothed with short, soft hairs.
Recurved: bent backward or downward.
Reflexed: abruptly recurved.
Reticulate: netted.
Revolute: rolled backwards.
Scabridulous: rough to the touch.
Sericeous: silky.
Serrate: saw-toothed, the teeth pointing forward.
Serrulate: minutely serrate.
Sessile: without a stalk.
Sinuate: with serpentine margins.
Stipel: a secondary stipule.
Stipitate: with a *stipe* or stalk.
Stipule: one of (usually) two appendages at the base of a petiole or leaf-stalk.
Striae: small elongate ridges.
Striate: bearing striae.
Sub-: prefix meaning 'almost' or 'somewhat', e.g. suborbicular, subglabrous, etc.

Subulate: awl-shaped.
Terete: rounded, not angled or channelled.
Tomentellous: thinly tomentose.
Tomentose: clothed with felted hairs or *tomentum*.
Tomentum: felted hairs.
Truncate: ending abruptly as if cut straight across.
Undulate: wavy.
Venation: system of veins or secondary nervation.
Ventrally: on the face, or towards the central axis; adaxially.
Villose: with long, soft, shaggy hairs.

INDEX

Abele .	24
Aspen .	25
Black Maul .	see 5
Grizette	see 5
Mottled Spaniards	see 5
Osier .	9
Poplar, Balsam .	28
Poplar, Black	26
Poplar, Grey	24 × 25
Poplar, Lombardy	see 26, 27b
Poplar, White	24
Populus alba L.	24
alba × tremula	24 × 25
angulata Ait. forma cordata (David ex Thomas) Rehder	see 27b
balsamifera L.	see 28
betulifolia Pursh	see 26
× canadensis Moench var. eugenei (Simon-Louis ex Mathieu)	see 27b
× canadensis Moench var. marilandica (Bosc ex Poir.) Rehder	27b
× canadensis Moench var. regenerata (Henry) Rehder	see 27b
× canadensis Moench var. serotina (Hartig) Rehder	27a
candicans Ait.	28
× canescens (Ait.) Sm.	24 × 25
deltoides × nigra	27a
deltoides Marsh.	see 27a
× *gileadensis* Rouleau	= 28
monilifera Ait.	see 27a
nigra L. .	26
ssp. betulifolia (Pursh) W. Wettst.	see 26
var. afghanica Aitch. et Hemsl.	see 26
var. betulifolia (Pursh) Torrey	see 26
var. italica Muenchh.	see 26

193

    var. plantierensis Schneid. . . . . see **27b**
    × robusta Schneid.. . . . . . see **27b**
Rhabdophaga rosaria . . . . . . see **14**
Salix acuminata Sm. . . . . . see **2b, 10**
    acutifolia Willd. **8**
    aegyptiaca L. . . . . . . see **10**
    alba L. var. alba **3**
        var. britzensis Spaeth . . . . see **3a**
        var. caerulea (Sm.) Sm. **3b**
        var. chermesina hort. . . . . see **3a**
        var. stenophylla . . . . . see **3b**
        var. vitellina (L.). Stokes **3a**
    alba × fragilis . . . . . . **3 × 2**
    alba × pentandra . . . . . . **3 × 1**
    alba var. vitellina × babylonica . . . **3a × 4**
    alba var. vitellina × fragilis . . . . **3a × 2**
    × alopecuroides Tausch . . . . see **2a**
    alpina Scop. . . . . . . see **21**
    × ambigua Ehrh. . . . . . **14 × 17**
    amygdalina L. . . . . = **5**
    andersoniana Sm. . . . . . see **15**
    aquatica Sm. . . . . . . = **13a**
    arbuscula L. **20**
    arenaria L. sec.Sm. . . . . . see **18**
    arenaria L. sec.Flod. . . . . see **17**
    argentea Sm. . . . . . . see **17**
    atrocinerea Brot. . . . . . = **13b**
    aurita L. **14**
    aurita var. minor Anderss. . . . . see **14**
    aurita var. nemorosa Anderss. subvar. virescens
        Anderss. . . . . . . see **14**
    aurita × cinerea . . . . . **14 × 13**
    aurita × cinerea ssp. oleifolia . . . see **13a**
    aurita × herbacea . . . . . **14 × 22**
    aurita × repens . . . . . **14 × 17**
    aurita × viminalis . . . . . **14 × 9**
    babylonica L. . . . . . see **3a × 4**
    babylonica salamonii Carrière . . . see **3a × 4**
    babylonica hybrids . . . . . see **3a**
    babylonica × fragilis . . . . . **4 × 2**
    basfordiana Scaling . . . . . see **3a × 2**
    bicolor Ehrh. . . . . . see **13 × 16**
    × blanda Anderss. . . . . . see **4 × 2**

breviserrata Flod.	see **21**
× calodendron Wimm.	**10**
caprea L.	**12**
ssp. *sericea* (Anderss.) Flod.	= **12a**
var. *alpina* Gaudin	= **12a**
var. *coaetanea* Hartm.	= **12a**
var. sphacelata (Sm.) Wahlenb.	**12a**
caprea × cinerea	**12** × **13**
caprea × viminalis	**12** × **9**
"cardinalis"	see **2c**
× cernua E. F. Linton	**22** × **17**
× charrieri Chass.	see **14** × **13**
× chouardii Chass. et Görz	see **13** × **9**
× *chrysocoma* Dode	= **3a** × **4**
cinerea L.	**13**
ssp. *atrocinerea* (Brot.) Silv. et Sobr.	= **13b**
ssp. cinerea	**13a**
ssp. oleifolia Macreight	**13b**
var. oleifolia Gaudin	see **13b**
cinerea × phylicifolia	**13** × **16**
cinerea × purpurea	**13** × **6**
? cinerea × purpurea × viminalis	?**13** × **6** × **9**
cinerea × repens	**13** × **17**
cinerea × viminalis	**13** × **9**
daphnoides Vill.	**7**
var. norvegica Ag.	see **7**
var. pomeranica (Willd.) Koch	see **8**
dasyclados Wimm.	see **10**
decipiens Hoffm.	see **2c**
× dichroa Doell	see **13** × **6**
× doniana Sm.	**6** × **17**
× ehrhartiana Sm.	**3** × **1**
elaeagnos Scop.	**11**
elegantissima C. Koch	see **4** × **2**
ferruginea G. Anders. ex Forbes	see **13** × **9**, **14** × **9**
foetida Schleich.	see **20**
× forbyana Sm.	?**13** × **6** × **9**
fragilior Host	= **2a**
fragilis L.	**2**
forma *latifolia* Anderss.	= **2a**
var. britannica Buchanan-White	see **2b**
var. decipiens (Hoffm.) Koch	**2c**

    var. fragilis **2**
    var. furcata Seringe ex Gaudin **2a**
    var. latifolia Anderss. . . . . . see **2**
    var. russelliana (Sm.) Koch **2b**
fragilis × pentandra . . . . . **2 × 1**
fragilis × triandra . . . . . see **2a, 2c**
× friesiana Anderss. . . . . . **17 × 9**
× fruticosa Doell . . . . . . **14 × 9**
fusca L. . . . . see **17**
× grahamii Borrer ex Baker . . . **14 × 22 × 17**
    var. moorei (F. B. White)
      Meikle . . . . . see **14 × 22 × 17**
helix L. . . . . see **6**
helvetica Vill. . . . . see **18**
herbacea L. **22**
herbacea × repens . . . . . . **22 × 17**
hibernica Rechinger f. . . . . . see **16**
hippophaifolia Thuill. . . . . . = **5 × 9**
hoffmanniana Sm. (1828) non Bluff et
  Fingerh. (1825) . . . . . . = **5a**
× holosericea Willd. . . . . see **12 × 9**
incana Schrank . . . . . . = **11**
lambertiana Sm. . . . . . . see **6**
lanata L. **19**
lanceolata Sm. . . . . . = **5 × 9 (b)**
lapponum L. **18**
× laurina Sm. . . . . **13 × 16**
macnabiana Macgillivray . . . see **14 × 22 × 17**
× margarita F. B. White . . . . **14 × 22**
× meyeriana Rostk. ex Willd. . . . **2 × 1**
× mollissima Hoffm. ex Elwert . . . **5 × 9**
    var. hippophaifolia (Thuill.) Wimm. . **5 × 9 (a)**
    var. mollissima . . . . see **5 × 9 (b)**
    var. undulata (Ehrh.) Wimm. . . **5 × 9**
mollissima Sm. . . . . . see **12 × 9**
× *moorei* F. B. White . . . = **14 × 22 × 17**
× multinervis Doell . . . . . **14 × 13**
myrsinifolia Salisb. **15**
myrsinites L. **21**
    var. arbutifolia Syme . . . . see **21**
    var. procumbens (Forbes) Syme . . see **21**
    var. serrata Syme . . . . . see **21**
nigricans Sm. . . . . . . = **15**

oleifolia Sm., non Vill. . . . . = **13b**
× pendulina Wenderoth . . . . **4 × 2**
  var. blanda Anderss. . . . see **4 × 2**
  var. elegantissima C. Koch . . see **4 × 2**
pentandra L. **1**
phylicifolia L. **16**
  var. lejocarpa Anderss. . . . see **16**
pontederana Willd. . . . . see **13 × 6**
purpurea L. **6**
  ssp. lambertiana (Sm.) A. Neumann ex
    Rechinger f. . . . . . . see **6**
  var. gracilis Gren. et Godr. . . . see **6**
purpurea × repens . . . . . . **6 × 17**
purpurea × viminalis . . . . . **6 × 9**
× reichardtii A. Kern. . . . . **12 × 13**
repens L. **17**
  var. argentea (Sm.) Wimm. et Grab. . . see **17**
  var. ericetorum Wimm. et Grab. . . see **17**
  var. fusca Wimm. et Grab. . . . see **17**
repens × viminalis . . . . . . **17 × 9**
reticulata L. **23**
rosmarinifolia L. . . . . . see **17**
× rubens Schrank . . . . . . **3 × 2**
× rubens nothovar. basfordiana (Scaling
  ex Salter) Meikle . . . . . **3a × 2**
  forma basfordiana Meikle . . . **3a × 2 (a)**
  forma sanguinea Meikle . . . **3a × 2 (b)**
× rubra Huds. . . . . . . **6 × 9**
sanguinea Scaling nom. nud. . . . see **3a × 2**
× sepulcralis Simonk. . . . . see **3a × 4**
× sepulcralis Simonk. nothovar. chrysocoma
  (Dode) Meikle . . . . . . **3a × 4**
× sericans Tausch ex A. Kern. . . . **12 × 9**
× smithiana Willd. . . . **13 × 9** (see **12 × 9**)
× smithiana nothovar. ferruginea (G. Anderss. ex
  Forbes) Leefe . . . . . see **13 × 9**
× sordida A. Kern. . . . . . **13 × 6**
sphacelata Sm. . . . . . = **12a**
stuartiana Sm. . . . . . . see **18**
× subsericea Doell . . . . . . **13 × 17**
tenuior Borrer . . . . . see **13 × 16**
trevirani Spreng. . . . . . = **5 × 9**
triandra L. **5**

var. hoffmanniana Bab.	**5a**
triandra × viminalis	**5 × 9**
× undulata Ehrh.	see **5 × 9 (b)**
viminalis L.	**9**
var. angustissima Coss. et Germ.	see **9**
var. intricata Leefe ex Bab.	see **9**
var. linearifolia Wimm. et Grab.	see **9**
var. stipularis Leefe ex Bab.	see **9**
violacea	see **8**
× viridis Fries	**3 × 2**
waldsteiniana Willd.	see **20**
wardiana Leefe ex F. B. White	see **13 × 16**
woolgariana Borrer	see **6**
Sallow, Gorgomel	see **2b, 10**
Sallow, Grey	**13a**
Sallow, Jargomel	see **10**
Sallow, Rusty	**13b**
Sarda	see **5**
Willow, Almond	**5**
Willow, Bay	**1**
Willow, Bedford	**2b**
Willow, Belgian Red	see **2c**
Willow, Crack	**2**
Willow, Creeping	**17**
Willow, Cricket-bat	**3b**
Willow, Dark-leaved	**15**
Willow, Downy	**18**
Willow, Dwarf	**22**
Willow, Eared	**14**
Willow, Goat	**12**
Willow, Golden	**3a**
Willow, Mountain	**20**
Willow, Net-leaved	**23**
Willow, Purple	**6**
Willow, Tea-leaved	**16**
Willow, Violet	**8**
Willow, White	**3**
Willow, White Welsh	**2c**
Willow, Whortle-leaved	**21**
Willow, Woolly	**19**
Yellow Dutch	see **5**